A WORLD NOT
MADE FOR US

SUNY SERIES IN ENVIRONMENTAL PHILOSOPHY AND ETHICS
J. Baird Callicott and John van Buren, editors

KEITH R. PETERSON

A World Not Made for Us

Topics in Critical Environmental Philosophy

Published by State University of New York Press, Albany
© 2020 State University of New York
All rights reserved

No part of this book may be used or reproduced in any manner whatsoever without written permission. No part of this book may be stored in a retrieval system or transmitted in any form or by any means including electronic, electrostatic, magnetic tape, mechanical, photocopying, recording, or otherwise without the prior permission in writing of the publisher.

For information, contact State University of New York Press, Albany, NY
www.sunypress.edu

Library of Congress Cataloging-in-Publication Data

Names: Peterson, Keith R., author.
Title: A world not made for us : topics in critical environmental philosophy / Keith R. Peterson.
Description: Albany : State University of New York Press, 2020. | Series: Suny series in environmental philosophy and ethics | Includes bibliographical references and index.
Identifiers: LCCN 2019048729 | ISBN 9781438479590 (hardcover) | ISBN 9781438479613 (ebook) | ISBN 9781438479606 (paperback)
Subjects: LCSH: Environmentalism—Philosophy. | Environmental ethics. | Philosophy of nature. | Nature—Effect of human beings on. | Human ecology.
Classification: LCC GE195 .P479 2020 | DDC 363.7001—dc23
LC record available at https://lccn.loc.gov/2019048729

10 9 8 7 6 5 4 3 2 1

For my parents

[The anthropocentric world view] corresponds to a certain dream image of the world, fondly framed by humankind at all times. It permits us to consider ourselves, in our capacity as a spiritual being, the crowning achievement of the world. In this manner we misunderstand not only the world but also our own being; and, rightly considered, this is not even to our advantage. Our task is *to come to terms with a world not made for us*—a far greater objective and one worthy of our power of self-determination.

—NICOLAI HARTMANN, *New Ways of Ontology*

CONTENTS

- ix   Preface
- xiii Acknowledgments
- 1   Introduction. Environmental Philosophy: Anthropocentrism, Intrinsic Value, and Worldview Clash

PART I
Anthropocentrism and Philosophical Anthropology

- 19   1 Anthropocentrism, Dualism, and Models of the Human
- 47   2 The Unfinished Animal

PART II
The Intrinsic Value of Nature

- 73   3 The Problem of Intrinsic Value and the Primacy of Priorities
- 97   4 Environmental Values and Vital Priorities
- 117 5 Political Ecology and Value Theory

PART III
Ecological Ontology

- 145 6 Metascientific Stances and Dependence
- 171 Conclusion. A World Not Made for Us

- 177 Notes
- 211 Works Cited
- 223 Index

PREFACE

This book attempts to define the core principle that should inform collective thinking about the global ecological crisis, namely, the recognition of human asymmetrical dependence upon the more-than-human world. The implications of recognizing this principle of dependence are explored in the domains of philosophical anthropology, value theory, and ontology. The recognition of human dependence—if consistently honored across epistemological, ontological, and sociocultural registers—reminds those of us who have forgotten that we live in *a world not made for us*.

The title of the book expresses a basic tenet of environmentalism, one that curiously runs against the grain of much contemporary philosophy. This tension has been reflected in my own experience. Growing up in a relatively rural part of northeastern Ohio, I spent a lot of time in the woods filled with the impression that I was merely a small, incompetent actor in a far larger and immensely more complex scenario. I am sure this is an experience shared by many North American environmentalists. Unlike most of them, however, later in life I also spent a lot of time reading many human exceptionalist German and French philosophers. For a relatively brief but especially formative span of years, enamored with German Idealism, Heidegger and Derrida, central Continental anti-realists, I thought that there was no bigger idea than that humans give to the world all the meaning that they could ever discover in it. I was gradually dissuaded from acceptance of this core humanist axiom by becoming further immersed in environmental philosophy and environmentalism. The allegedly naïve feeling and experience of the independence and indifference of the more-than-human natural world to my wishes and dreams that had first come to me in the woods of my childhood unassumingly returned to its former prominence in my psyche. I consider it one of the greatest disservices of contemporary philosophy to have convinced many theorists to deny, ignore, or background this experience.

The path to the position presented in this book was just as idiosyncratic as the view it ultimately presents. Initially, the philosophy I studied trained me to be highly wary of claims about "human nature," especially in regards to any forms of essentialism or determinism; yet the more I studied environmentalism, the clearer it became that some extended reflection on what was once called "philosophical anthropology" was needed to make sense of the ecological crisis. My search for intellectual allies led me to American philosopher Marjorie Grene, who introduced me to the more obscure German philosophical anthropologists whom, despite my Continental training, I had never read. While the social and political contexts of interwar Germany and late twentieth-century North America differed dramatically, thinkers from both these periods were wrestling with, and struggling to overcome, traditional, dualistically conceived categories. As is made clear throughout the book, Australian ecofeminist Val Plumwood's sophisticated critique of hegemonic dualisms played a large part in my attempts to bring humanistic and environmentalist traditions together. After researching and teaching courses on these thinkers for years, I found a way to integrate what appeared to be their otherwise unrelated projects.

Compared to the anthropological theories that impacted my understanding of environmentalism, integrating the idea that in order to pursue environmental social change we cannot avoid "values" talk was significantly more difficult given my philosophical training. Like most of my peers in graduate school, I was a good Nietzschean, congenitally suspicious of any talk of values. In an era that touted "family values" (Which ones? Whose family?), I reluctantly surrendered to the realization that values talk is not only useful in approaching environmental discourses, but absolutely essential if what environmentalists aim to do is generate social change. I started wrangling with the environmental ethicists' distinction between instrumental and intrinsic value, but quickly realized that this dichotomy was of little use for understanding environmental values, as well as for thinking about the role or nature of values in ethics more broadly. On my way to becoming a more pluralistic Continentalist and committed environmental philosopher, I discovered Max Scheler's remarkable book on ethics, *Formalism*, and soon after Nicolai Hartmann's version of value ethics, which provided me with almost all of the questions and at least some of the answers I needed to begin to think differently about environmental value theory. I realized that if social change is what environmentalists want, they must articulate a cluster of values distinct from those instituted by the existing hegemonic culture, and they must explain why it is better for both humans and nonhumans to adopt this cluster of values. More specifically, I came to see that there is a link between value articulation and practice, between values and action, that needed to be reconsidered at much greater depth than had been done in existing discourses. Given the less than transparent relationship between values and actions, I came to see that the best that we can expect from ethical theorizing is some guidance through the thicket of moral conflicts and wicked problems that face us on a daily basis.

Yet more than merely offering some modicum of moral guidance, I came to realize that values talk also provides a very powerful way to synthesize otherwise divergent programs for responding to the crisis within environmentalism, an insight gleaned from my encounter with the work of ecosocialist Joel Kovel and ecocommunitarian anarchist John P. Clark. The impact of their ideas on this work will be evident in later chapters. Finally, my path to the realist, ecological materialist ontology and "metascientific stance" defined in this book was made significantly smoother by virtue of a timely encounter with recent realist trends in contemporary Continental philosophy, on the one hand, and more recent research in science studies on the other, both of which helped me to give a new spin to an older and long-forgotten body of work.

I hope that the distinctive way I approach these topics in this book allows environmental philosophers and environmentalists to better pinpoint and correct some of the weaknesses with current discourses about the environment. If it helps readers to consider more thoughtfully the conditions under which we must create our future lives together and what those best lives might look like, all the better.

ACKNOWLEDGMENTS

Although some of this material has appeared in published form, almost all of it has been so thoroughly revised that it bears very little resemblance to the original. The following list of sources from which parts of the book might have sprung is thus of dubious utility for readers, but might satisfy the curiosity of a few as much as it does the conventions of academic publishing:

"Ecosystem Services, Nonhuman Agencies, and Diffuse Dependence,"
    Environmental Philosophy 9, no. 2 (Fall 2012): 1–19.
"Bringing Values Down to Earth: Max Scheler and Environmental Philosophy,"
    *Appraisal: The Journal of the Society for Post-Critical and Personalist Studies,
    Re-Appraisal: Max Scheler (Pt 2)* 8, no. 4 (October 2011): 3–12.
"All That We Are: Philosophical Anthropology and Ecophilosophy," *Cosmos and
    History: The Journal of Natural and Social Philosophy* 6, no. 1 (2010): 60–82.
"From Ecological Politics to Intrinsic Value: An Examination of Kovel's Value
    Theory," *Capitalism Nature Socialism* 21, no. 3 (2010): 81–101.
"Stratification, Dependence, and Nonanthropocentrism: Nicolai Hartmann's Critical
    Ontology," in *Ontologies of Nature*, edited by Gerard Kuperus and Marjolein
    Oele, 159–80. New York: Springer Berlin Heidelberg, 2017.

Thanks are due to the publishers of these pieces. The Hartmann epigraph above (translation modified and my emphasis) is from his *New Ways of Ontology*, trans. R. Kuhn (Chicago, IL: Henry Regnery, 1953), 90.

This book has been in progress for many years. Much of this work has drawn on presentations given at a number of venues, including the International Association for Environmental Philosophy (IAEP), International Society for Environmental Ethics (ISEE), Society for Phenomenology and Existential

Philosophy (SPEP), the Max Scheler Society of North America, Nicolai Hartmann Society (NHS), and the Society for Phenomenology and the Human Sciences (SPHS), among others. I thank these organizations for providing a place for public discussion of the work. Thanks go to Leah Zuo at Bowdoin College for inviting me to participate in the Kemp Symposium in 2014, where parts of what became chapter 6 were initially discussed. I also thank Colby College for the funding to participate in all of these events, and for the generous one-year sabbatical during which the first draft of the book was written. One particularly tough but generous anonymous reviewer forced me to cut down an otherwise unwieldy and more diffuse manuscript, and I appreciate the extensive comments and recommendations for change that reader provided. Rick Elmore selflessly devoted much of his time and critical acumen to reading a large portion of a very early draft of the book, and it undoubtedly improved as a result. Some students, especially Ryan Lam and Erin Maidman, also contributed valuable feedback on the manuscript as it was in its final stages. I would also like to thank John P. Clark both for being an inspiring exemplar and for his insightful and profuse comments on the manuscript. I am of course solely accountable for whatever flaws remain.

This book would not have been possible without the unconditional support of my family. I am immeasurably grateful to ÖDT, bighearted and fierce; to EHP and NRP, the best kids ever and wise beyond their years; and to my parents, whose quiet, unequivocal support has for years sustained my peculiar endeavors.

INTRODUCTION

Environmental Philosophy

Anthropocentrism, Intrinsic Value, and Worldview Clash

ENVIRONMENTAL PHILOSOPHY HAS CHALLENGED THE DOMINANT WESTERN CULTURE'S CONCEPTION OF human nature through critiques of anthropocentrism (human chauvinism). It has annoyed the mainstream with critiques of instrumental rationality and its plea on behalf of the intrinsic value of nature. It has irritated nonenvironmentalists and even some environmentalists with its criticism of mechanism or the reductionist scientific worldview and has argued in favor of some form of ecological worldview. The *critique of anthropocentrism*, the *intrinsic value of nature*, and the *ecological worldview* are central topics for environmental philosophers, appearing across a wide range of environmentalist writing, from environmental ethics and policy to political ecology, ecocriticism, and metaphysics. As I understand them, these topics have characterized environmental philosophy since its inception in the 1970s.

In the widespread environmental imaginary of a few decades ago, perhaps *the* central term of engagement for environmental philosophers and ethicists was the concept of anthropocentrism. Anthropocentrism—whose core meaning is human-centric evaluation—was also considered by many to be one of the central causes of the environmental crisis. Identifying its historical and conceptual sources and pulling them out by the roots formed a large part of the environmentalist response.[1] By the early 1990s, the Australian environmental philosopher Warwick Fox could write that "virtually every paper and book that ecophilosophers have written either implicitly or explicitly develops some kind of answer to [the] question 'what's wrong with being anthropocentric?'"[2] It effectively encapsulates a number of issues that have attracted critical attention: the Modern western dualistic opposition between humans and nonhuman nature; the notion of human chauvinism or human-centric evaluation; and the concept of nature as mere resource passively awaiting instrumentalist exploitation. Current debates around the concept of the Anthropocene suggest that renewed attention to this topic is warranted.

In addition to the critique of anthropocentrism, a "new ethics" was called for by many environmental philosophers. Are traditional ethical categories and theories so fundamentally anthropocentric that a completely new ethics is required? Adopting a nonanthropocentric perspective would mean accepting the propositions that nonhumans have moral worth, and that they must be taken seriously in human decisions about environmental issues. In a nonanthropocentric ethics, this also means that in cases of conflict their interests may often carry greater weight than those of humans. Environmental ethics might have to be "new" if traditional theories cannot accommodate these points.[3] After briefly entertaining the possibility of using the existing concepts of "rights" or "standing" for nonhumans in the 1970s, environmental ethics came more and more to be identified with arguments for the intrinsic value of nature. A trickle of references to the intrinsic value of nature in the 1970s gradually became a steady stream in the late 1980s, and the high-water mark was reached in the debate in the 1990s.[4] Finding the appropriate epistemological, ontological, and normative arguments to secure the concept became a major preoccupation. J. Baird Callicott explicitly declared that the distinctive feature of environmental ethics would be its claim that nature possesses intrinsic value.[5] He claimed that "the most important philosophical task for environmental ethics is the development of a non-anthropocentric value theory," and he defined the difference between anthropocentric and nonanthropocentric ethics in terms of intrinsic value.

> An anthropocentric value theory (or axiology), by common consensus, confers intrinsic value on human beings and regards all other things, including other forms of life, as being only instrumentally valuable, i.e., valuable only to the extent that they are means or instruments which may

serve human beings. A non-anthropocentric value theory (or axiology), on the other hand, would confer intrinsic value on some non-human beings.[6]

Reflecting on the concept itself, some writers noted that this very specific quest for the establishment of the intrinsic value of nonhuman nature was motivated by the need to identify some transcultural anchor of environmental value against the backdrop of value relativism. It is only if some value "independent of and overrid[ing] individual human judgment and ... relative and evolving cultural ideals" could be found that environmental value would be safe from provincial nature-exploitative interests.[7] Although it was and remains a fundamental part of the discourse of environmental ethics, I will provide some reasons to doubt the efficacy of this approach to value theory in part two of the book.

Finally, by the 1980s it became conventional among environmental philosophers to contrast an "ecological worldview" with "the Modern scientific worldview"—where the latter is taken to be an expression of Cartesian dualism, atomism, mechanism, and reductionist materialism—and to indict it as one of the central causes of the ecological crisis. The theoretical and technological transformations characterizing the Scientific Revolution, along with its supporting Judeo-Christian tradition, were seen as chief contributors to the highly anthropocentric, exploitative relationship of humankind to nature in western culture. From Arne Naess's contrast between thing- and field-ontology (1972), to Carolyn Merchant's case against Modern science and her plea for a return to a holistic, organismic conception of nature (1980), to Charles Birch and John Cobb Jr.'s mechanistic and ecological models of the living (1981), to J. Baird Callicott's "metaphysical implications of ecology" (1986), and, ultimately, to the elaboration of these alternative conceptions by other writers during the 1990s, including Warwick Fox (1990, 1995), Bryan Norton (1991), Murray Bookchin (1996), and Arran Gare (1996), this contrast became a defining feature of environmental philosophy. Since "worldview" talk is also central to the post-Kantian tradition, the Continentalists among environmentalists seamlessly extended the general antipathy to the sciences in the dominant strains of Continental philosophy into environmentalism, and works like Neil Evernden's (1985) and David Abram's (1995) also traded on a series of oppositions central to the grand contrast between mechanist and ecological worldviews. Even today, there are calls for "worldview remediation" and proposals to explicitly use the worldview concept as a tool in sustainable development debates.[8]

Although the figures and approaches listed are often conceived as antagonistic to one another (e.g., deep ecology is often not compatible with pragmatism, nor is ecophenomenology compatible with social ecology), they share the preoccupation with distinguishing a minority environmentalist "ecological worldview" from a hegemonic "mechanistic worldview." I will call this the "worldview clash" model for thinking about science-environmentalism relations.

This book is organized around these three major topics. Used in the sense of "central issues" or "places" (topoi) of contention and thought, topics are "clear enough and serious enough to engage a mind to whom they are new, and also abrasive enough to strike sparks off those who have been thinking about these things for years."[9] This book directly engages with the fundamental assumptions, categories, concepts, and value priorities that characterize large parts of environmentalist thinking, and considers the conditions under which environmentalists and others generally think about the nature of humankind (philosophical anthropology), how they think about the value of nonhuman nature (metaethics and value theory), and how they understand more-than-human nature generally (ontology and epistemology).[10] The three parts of the book deal with these three broad topics. I have organized the book in terms of them not because I think they embody timeless philosophical questions, but because initially I found it helpful to organize the wide array of literature that falls under the heading of environmental philosophy in this way, and hopefully it will be for others. For introducing environmental philosophy to those unfamiliar with it, they also serve a heuristic function, like a ladder to be pulled up and carefully dismantled once one reaches the desired height.

I consider environmental philosophy to be an informed examination of the concepts, categories, assumptions, and priorities in historically and culturally diverse human interaction with the human and nonhuman natural world, along with the implications of their mostly tacit operation. Philosophers have long recognized that much human activity is caused and conditioned in large part—but never exclusively—by linguistic and conceptual categories, value priorities, and unspoken assumptions that remain mostly invisible to those who think and act with them. Philosophers are particularly good at thinking about such conditions, and if they have shown that these conditions motivate anti-environmental activity in significant ways, finding the flaws in these frameworks and correcting them ought to play a role in generating the kind of social change environmentalists desire. This definition of environmental philosophy already implies that the scope of such philosophical work is far broader than most people usually think. Contrary to popular belief, philosophy is not merely about policing argumentation by day and soul-searching by night. The scope of environmental philosophy encompasses the most fundamental questions of human experience. What is "nature"? Who are "we," and what is the place of human beings in nature? What is the good life for the individual, and for the human and more-than-human community? Environmental philosophy has never been exclusively philosophy about the environment because the questions it raises cover most major philosophical disciplines, including epistemology, ontology, ethics, aesthetics, and political philosophy. How do communities know they have environmental problems? How do they divide and categorize the social and natural worlds when they frame environmental problems? Do ecosystems and the land have a value, integrity, or beauty that ought to be preserved? If communities can agree that valuing nonhuman nature in certain ways is important, how do they negotiate and institute the necessary social changes? What kinds of social institutions are essential for creating sustainable societies?

Even if examining tacit assumptions is important work, and if the scope of environmental philosophy is wider than believed, why should anyone who cares about the environment care about this kind of work? Won't thoughtful policymakers and their scientific consultants eventually arrive at the best scientific and democratic solutions to our environmental problems? Unfortunately, this is not very likely. This is because without serious thought given to the traditional models and frameworks used to characterize problems in the first place, their "solutions" will continue to perpetuate (with only minor modifications) the same harmful conceptual frameworks that have led to the current situation. Reflective environmentalists should care about this work because the explanations that philosophers and environmentalists have given of the causes of the crisis—whether these lie deep in "human nature," mechanistic science, industrialization, capitalism, human supremacist religions, and so forth—have been at least partially right. But these explanations are so little known and remain so contrary to current lifeways of the world's economically and politically dominant societies that they cannot penetrate mainstream thinking. Worse, the environmentalist or philosophical explanations themselves often entail assumptions that prevent them from achieving the kind of environmentalist social change they hope to accomplish. They may even create obstacles to it. In light of this, one of the primary tasks of this book is to review those explanations, assess their strengths and weaknesses, and improve on them so that environmentalist social change seems feasible and imaginable, rather than remaining impossible and unimaginable.

DUALISMS AND CRITICAL ENVIRONMENTAL PHILOSOPHY

The position taken in this book agrees with that of the late ecofeminist philosopher Val Plumwood in its claim that each of the three problems sketched above is rooted in the dominant dualist conceptual framework of western philosophy, what has been called the "logic of domination." Since I will explain the operation of this dualist logic more extensively in the first chapter, I only summarize it here. Plumwood herself examined dualistic thinking as it relates to feminist theory, environmentalism, colonialism, and many other areas of philosophical debate. Some of the dualistically conceived categories she discussed include culture (mind, reason) and nature, mind and body, male and female, form and matter, reason and emotion, freedom and necessity, human and (nonhuman) nature, production and reproduction, mental and manual, public and private, civilized and primitive, subject and object, and self and other. She noted that various liberation struggles have had to engage with these implicit or explicit dualisms of western culture: feminism with masculine and feminine, racism and anti-colonialism with civilized and primitive, classism with mental and manual, and environmentalism with human (mind, reason, culture) and nature. In all of these forms of oppressive dualizing she identified a logical pattern of "hegemonic centrism" that conditions thought and has five characteristics. Firstly, the terms of the dualism are "hyperseparated," or treated as radically exclusive disjuncts. Not only are the two poles taken to be different in kind, but the "different" is conceived as inferior from the point of view of the "center." This applies to each pair of terms in the examples listed above, where the first is conventionally construed as superior and the second inferior. Secondly, hyperseparation works in tandem with "homogenization" of the terms, where every member of the class is (usually wrongly) considered to possess all of the characteristics of every other member. "Man" is opposed to "animal," as if there were no relevant intragroup differences in either class. Thirdly, the second term is always "backgrounded," or its value (as that on which the superior term depends) is actively denied and taken for granted by the first, and considered to be inessential. For instance, women's domestic labor is invisible, undervalued, and taken for granted in most economic calculations. Fourthly, the second term is "assimilated" to the first, in that it is defined negatively in relation to it. For example, if humans are rational then nonhumans are nonrational, rather than positively defined in their own terms. Finally, given all of the above, the second term is normally considered to be a means to the ends of the superior term, it is treated as "instrumentally valuable" only, without any intrinsic worth or ends of its own. Plumwood argued that anthropocentrism, instrumentalism, and mechanism have to be seen as expressions of this widespread western dualizing logic of domination that reaches back to Greek philosophy.[11] They are comprehensive problems that require a comprehensive systematic response.

The point of highlighting this dualist logic is to claim that the way the classical problems of environmental philosophy are framed depends directly on it, and this means that the supposed solutions to them usually depend directly on it as well. Instead of identifying the logic and undermining it as a whole, many responses simply react to certain limited aspects of it, and this seriously dampens the efficacy of their engagement with the environmental crisis. While I criticize some of these earlier responses in what follows, in the interest of space I spend less time on this and more time on articulating a positive position. The positive position developed throughout this book is articulated in response to various forms of this classical dualism. The task is to generate a critical environmental philosophy that unmasks the function of this logic in the domains of anthropology, value theory, and ontology, and engages in multiple intersecting strategies of responding to these dualisms. While still attending to the different features of the logic Plumwood identified, I will emphasize the feature of dependence denial and backgrounding, and will encapsulate the response to this logic in terms of a "principle of dependence." In plain terms, it states that the asymmetrical dependence of human on more-than-human nature has always been at the heart of environmentalist concern. Human dependence on nonhuman nature is the most chronic, indispensable, and palpable experience of more-than-human nature—we literally live and breathe it. However, acknowledging human dependence on nonhuman processes or systems demands a more fundamental rethinking of categories than traditional approaches engage in, since it challenges more deeply the dualistic conceptual framework in terms of which the problems are framed. If the historically dominant anthropocentric view has been to regard human beings as independent of nature, for instance, then an environmental philosophy is a philosophy that asserts the asymmetrical dependence of human life in its physical, biological, psychological, and cultural registers on the living and nonliving environment, rather than its independence from it. Dependence, in addition, should be regarded as in itself plural, since it has multiple meanings in the contexts of anthropology, value theory, and ontology. Making the category of dependence a guiding thread for the treatment of the three core topics of environmental philosophy will help to correct deficient statements of the problems, and in theory reorient environmentalist social engagement.

Let me briefly point out some of the ways in which dualisms are manifest in the traditional statements of the three problems introduced above. In terms of anthropocentrism critique, the stereotypical Western conception of the environmental crisis (i.e., the Western "ecological imaginary") involves a basic dualism in which Humans on one side square off against nonhuman Nature on the other, and Humans do something terrible to Nature. The use of capital letters here reflects the traditionally homogenizing and universalizing features of dualistic thinking. This dualism is reflected in early responses to the crisis by environmentalists (and some philosophers). The fifth tenet of the popular deep ecology platform is a prime example: "Present human interference with the nonhuman world is excessive, and the situation is rapidly worsening."[12] Far from dismantling the dualism between humans and nature that undergirded anthropocentrism, early responses often reinforced it. Given the social location of many early theorists, the ecological imaginary especially foregrounded the value of forests, mountains, wildlife, and wilderness untouched by humans—that is, what was considered maximally nonhuman—in contrast with the human.[13] The conceptual polarity of human and nature thus merged with a specifically North American (and Australian) political geography that separated "wild," nonhuman nature and "civilization" in space as well as in thought. On this model, cows and cornfields were no more nature than were oil fields and coal mines.[14] Given this particular environmental imaginary, only certain problems appear to count as environmental, and certain aspects of the world as natural. This dualism was also reinforced in another way. For wilderness environmentalism, to be against anthropocentrism and for the environment meant avoiding all arguments regarding the value of nature that traced the origins of environmental value back to the human in any way. This meant that social ecologists and ecofeminists, who view practices of valuing or devaluing nature as direct reflections of human social relations, did not quite count as environmentalists. However, their work is indispensable for more comprehensive thinking about the environmental crisis as a social crisis with social conditions, and it undermines the analytical segregation of ethics from culture and social institutions. It is now more widely recognized that questions about human evaluations of nature must directly confront the social institutional and natural conditions of moral agency and social engagement, as well as conceptions of the natural world in which human agents are embedded. Empirically oriented social scientific political ecologists have also argued that speculative conceptions of human nature play very little role in the crisis, and that what is needed is more attention to power-infused social relations in concrete bioregional, natural-cultural places.[15] This is an important point, but it does not put an end to the need to think philosophically and conceptually about environmentalism and anthropocentrism since the conceptual frameworks widely employed to address environmental issues are also often left uninterrogated by these authors.

What is already evident in light of the dualist logic, however, is the universalizing tendency in the critique of anthropocentrism itself. *Humanity, humankind, man, Homo sapiens, the human species*, and *the human enterprise* all seem to denote a homogenous class with homogeneous interests. It is as if there is just one place for humans in the cosmos, and the anthropocentrism of nature exploiters implies that the privileged species is one and homogeneous. Social ecologists, ecofeminists, environmental justice theorists, and social scientific political ecologists have amply demonstrated that such a universal environmentalism artificially homogenizes a heterogeneous collection of human genders, classes, races, cultures, and communities, thereby rendering invisible the differential actions of particular groups of humans and their differential environmental impacts in particular bioregional locations around the globe. According to a historical materialist critique of early philosophical environmental theory, for example, it is really only some small part of humanity that is to be blamed for the environmental crisis, namely, the rich and powerful steering the capitalist juggernaut, rather than humans in general.[16] Many universalizing environmentalists (including some philosophers), have employed anthropocentrism critique in the undifferentiated, homogenizing sense, leading to sweeping claims about how humans in general are destroying the planet, supported by a species-concept of the human, when, in fact, it is specific human groups facilitated by powerful global institutions that have wreaked most of the destruction. The influential essay by historian Lynn White Jr. notes that "the impact of our race upon the environment has so increased in force that it has changed in essence," and echoes of this kind of universalism are evident in recent Anthropocene discourse.[17] While we should preserve anthropocentrism critique because of the conceptual issues it opens up, a critical environmental philosophy must "dehomogenize" within the class *humanity* in order to emphasize social, political, and environmental differences. The class *animal* or *nonhuman* must be symmetrically dehomogenized. The impulse to universalize, while perhaps well intentioned, is a response framed by the dualist conceptual framework that operates tacitly in most environmentalist and nonenvironmentalist thinking and practice.

Dualisms framing the topic of intrinsic value are just as obvious and just as detrimental to the development of environmental value theory. In light of Modern epistemology and ontology, values are considered subjective, relative, changeable features of human perception or judgment about the world, or at best features of cultural worldviews, and the world or nature is understood as originally a valueless domain of material in motion. Against this backdrop of the Modern constitution, the burden of proof seems to fall on environmental ethicists to show that objective intrinsic value exists and is possessed by nonhumans. This compels many writers to continue to sharply distinguish between the objective and subjective domains, rationality and emotion, and to engage in the quest for an objective, invariable, morally relevant universal property in the hopes of settling disputes between individually and culturally relative conceptions of nature's value. On the dualistic view, culture is coded *subjective* and *relative*, while a properly objective value would be *acultural* or *transcultural* and *transhistorical*. The anthropology implicit in such a conception stems from the eighteenth century, where universal reason reveals the truly timeless and objective truth of things. The metaethical problem concerning the existence of intrinsic value becomes central to the discourse. I believe that there is an experience of moral conflict over more-than-human nature's value expressed in the problem of relativism to which this debate implicitly and explicitly refers, but that experience is falsified when it is placed into the straightjacket of the Modern dualism opposing the objectivity and subjectivity of value, and its dualistically conceived intrinsic or instrumental character.

Finally, among the chief problems with the clash of worldviews model is the fact that even contemporary ecological science is often uncritically identified with an eighteenth-century Modernist conception of mechanistic science. This keeps many traditional dualized terms—such as life and matter, whole and part, mind and body—firmly in place while simply inverting their Modernist hierarchical relation. This also leaves the science of ecology in the rather peculiar position of being treated sometimes as one more extension of mechanism—and as thus useless or even harmful for environmentalism—and at other times as somehow fundamentally different from every other science—and so as ally and even justification for environmentalism. A similar point was made by philosopher of science Kristin Schrader-Frechette and taken up by later authors who call out environmentalists for not adequately recognizing the difference between "hard" or "scientific" ecology, and "soft" or "Romantic-political-metaphysical" ecology.[18] An important advantage of her distinction is that contemporary scientific ecology and metaphysical ecology (or "ecological worldview") may be treated separately, and it puts into question the ready identification of scientific ecology with eighteenth-century metaphysical mechanism. This simple shift toward considering different varieties of ecology already deflates the often-exaggerated dualistic framing of the clash of worldviews, and creates an opening into which more nuanced understandings of ecology and environmentalism may be inserted. For these more nuanced approaches, there is a struggle within the sciences (as well as without) for more politically sensitive, socially engaged sciences, in contrast to positivist, "value-neutral" science that has often easily ended up as legitimation for capitalist exploitation of more-than-human (and human) nature. From this standpoint, the problem is not so much "mechanism"—although the ontological and epistemological principles involved in it are worth examining—but an epistemology that prevents the recognition of the value-saturated interests driving the production of knowledge, as well as recognition of the many other social factors conditioning knowledge making in complex societies. This book adopts what I will call a "metascientific stance" that aims to see practicing scientists as socially engaged knowledge-producing agents embedded in their social contexts, and brings the tools of the history, philosophy, and social studies of science to bear in their analyses. By *metascientific stance* I mean the tacit assumptions about the nature, practices, goals, and place of the sciences in society. The relation between a metascientific stance and an articulated philosophy of science is analogous to the relation between a set of metaethical assumptions and an articulated normative ethical theory. By calling it a *stance* I acknowledge its irreducibly evaluative nature. It is unlikely that anyone would claim that a distinct philosophy of science belongs to environmental philosophy, but this does not mean that it does not often have very definite ideas about the nature and goals of the sciences. Thus, I will claim that the "clash of worldviews" model is a metascientific stance that often accuses capital-S "Science" of mechanism, instrumentalization, and domination of nature. This stance, unfortunately, throws the baby out with the bathwater. Unless one is willing to sacrifice the cognitive authority of the climate sciences in claims about global climate disruption, for example, or the authority of ecology in claims about biodiversity loss, a perspective on the sciences that does not consider scientific knowledge (merely) a worldview is absolutely essential for environmentalists. Critical environmental philosophy should engage with recent philosophy and social studies of the sciences in order to develop a more sophisticated metascientific stance toward the environmental sciences and the role of scientists and other environmental professionals in society. It should not fall prey to the naïve "science wars" opposition between classical scientific realism versus postmodern relativism, which is so clearly one of the most decadent expressions of traditional dualism. This means that the rational-social dualism in studies of knowledge making has to be overcome for critical environmentalism just as much as does the mind-body or human-nature dualism.

An environmental philosophy is "critical" if it explicitly takes into account and dismantles the dualistic conditions under which these problems are framed. I call the comprehensive response to the dualistic construal of these problems *critical environmental philosophy* or, in line with the definition that follows, *political ecology*. While the term *political ecology* has been used in many ways (discussed further in chapter 5), here I employ it to indicate three things. Anthropologically, it entails an embodied and embedded conception of the human, or ecological materialism. It takes human ontogeny (or developmental life span and its processes) seriously, and rejects dualistic and reductionist conceptions of humankind that often ignore it. While materialism or naturalism has always recognized the asymmetrical dependence of humankind on the physical world, this book adopts a nonreductive naturalism about human being. In terms of value theory, political ecology means recognizing the embeddedness of ethical relations within a larger context of social relations and institutions, often backgrounded by philosophers. The dependence denial that is a large component of the environmental crisis is not just an ethical problem, it is a social problem. Denial of social dependencies is itself a symptom of this deeper problem. Inspired by Murray Bookchin, John P. Clark, Plumwood and other ecofeminists, Joel Kovel and the ecosocialists, critical environmental philosophers have to be able to situate environmental ethics within in a larger social world. This includes better understanding the nature of values, the role of social ethos in value prioritization, and the role of ideologies and institutions in stabilizing this ethos. Finally, in relation to ecological worldview, the political ecology espoused here is informed by ecologist and social theorist Peter J. Taylor, social science political ecology, and categorial ontology, and develops a metascientific stance that recognizes structural ontological dependencies in the real world and dissolves the rational-social dualism in accounts of scientific knowledge production about environmental problems. The title of the book, *A World Not Made for Us*, is meant to provoke reflection not only on what *world* or nature means in environmentalist discourse, but also on the "us" for whom the world or nature ostensibly serves as resource, succor, essential life condition, or dumping ground. In the anthropocentric humanist tradition, "we" are "civilized," European, masculine, Christian, and (mostly) capitalist rational moral agents superior to all else on Earth. If, as environmentalists argue, the world is not made for this "us," but is the supporting and fecund home for all life on Earth, we have to find new ways of shaping and establishing a nonanthropocentric, nondualistic, nondominating human and nonhuman collective yet to come.

STRUCTURE OF THE BOOK

This brief review of the way that the three classical topics of environmental philosophy have been articulated reveals some important points. First, although environmental philosophers are generally unanimous in claiming that the traditional western view has been damagingly anthropocentric, there has been little unanimity about the conception of the human being that ought to inform a genuinely nonanthropocentric philosophy. This state of affairs calls for a review of the existing options, and for an assessment that is guided by the core principle of dependence and the rejection of dualist logic. Secondly, while many environmentalists have decried the instrumentalization, exploitation, or human domination of nonhuman nature, there is still little agreement over how nonexploitative environmental values are to be conceived. The once central debates over intrinsic value have moved to the periphery, and many other options are now on the table, including weak anthropocentric, pragmatic, ecofeminist, and virtue ethical value theories. All of these positions metaethically imply a certain conception of the valuing agent. A value theory guided by the insight into complex dependencies will be explored that is expressly nonsubjectivist, since the subjectivism of value in ethics, economics, and other fields is itself a key feature of the Modern dualistic constitution of anthropocentrism. Finally, while the critique of the mechanistic worldview has also been a frequent touchstone for environmentalists, here too responses to it have been, not surprisingly, diverse. The views taken on this topic adopt metascientific stances which situate the sciences relative to society and social environmental engagement in specific ways. In light of the principle of dependence and the rejection of dualism, I assess the mostly implicit metascientific options that have been offered by environmentalist philosophers as well as, to a lesser extent, the ontological frameworks for social and natural life employed by them. A pluralist, stratified ontology is presented as a response to dualism that enables clear articulation of human asymmetrical dependence on nonhuman nature, one that supports a more sophisticated metascientific stance for environmentalists beyond the clash of worldviews and science-driven environmentalism.

The three parts of the book deal with the three topics of anthropocentrism, intrinsic value, and ecological worldview. In part 1, the problem of anthropocentrism is further characterized in light of the dualistic framework, and Plumwood's critique of this logic is presented in greater detail. A quick review of typical environmentalist (rationalist and naturalistic) anthropologies follows, and it is argued that the categorial frameworks through which they are articulated fail to overcome dualism and adhere to the principle of dependence. In chapter two, a more coherent nonreductive, nondualistic "ecological materialist" conception of humankind as persons embodied and embedded in natural and social environments is presented in order to motivate better responses to dualism, and to provide a metaethics that anchors value perception in the naturalistic anthropological principles of plastic and surplus impulses, affective embodied cognition, and the structure of human action. Drawing on a variety of philosophical and empirical resources—from the classical German tradition of philosophical anthropology to feminism and contemporary developmental and cultural psychology—it outlines an ecological materialist anthropology that fully acknowledges human dependence on nonhuman nature in epistemic, ethical, and ontological registers.[19]

The chapters of part 2 begin with the debates over intrinsic value in order to show that the ways in which the problem of environmental value was framed continue to persist and hamper discussions of value and social engagement. Questions of value were usually asked within a rationalist framework that privileged the search for necessary and sufficient conditions for value ascription, rather than in terms of an exploration of the plural values humans experience and the many ways in which they are and may be prioritized. Although the well-intentioned search for proofs of the existence of intrinsic value of nature were meant to give us something that might serve as an obstacle to the instrumentalization of nature, they generally ignored the moral psychology of the agents who would be acting on recognition of intrinsic value in nature. In addition to the questionable assumptions implied in the definitions of instrumental and intrinsic value, early work typically underemphasized the sociocultural embeddedness of moral agents. Some value theories espoused by theorists are reviewed in an attempt to rebuild the framework for considering environmental value beyond dualism and in light of dependence. I introduce a novel, pluralistic value theory for environmentalism that takes its point of departure from both the social environmentalist writers and the tradition of axiological ethics. It holds that the experience of value is inherently plural, that moral and social life is life in the midst of never-ending conflicts of value, and to resolve them we resort to mostly unconscious patterned prioritization of values in social context. Environmentalism thus requires a value theory that can explain not only how prioritization of values concretely operates for individuals, but also how values are ordered within the social ethos of a given community and how different patterns of prioritizing come into conflict. Reframing environmental conflicts in terms of value priorities also leads more naturally to a social-deliberative, bottom-up rather than top-down model of environmental engagement. It has the potential to fuel an ecocollectivist response to the crisis in place of the oft-proposed social engineering of an elite policy-making class. Thus, making this shift is not only relevant to ethicists, but to anyone considering the values of nature for humans and nonhumans, including policy makers, political ecologists resisting the commodification of nature, anthropologists, and anyone involved in environmentalist knowledge-making and social engagement.

Part 3 of the book advocates adoption of a metascientific stance that differs from both the worldview clash model and positivist science-driven environmentalism. A virtue of political ecology is that it casts knowledge producers in society as socially engaged agents, both intervening in and responding to the "unruly complexity" of the intersecting processes through which knowledge of phenomena such as environmental problems is produced. The ontology implicit in these social studies of the sciences avoids many dualisms, but at the expense of obscuring macroscale relations of dependence. This largely relational framework is contrasted with this book's stratified one in which the claim that humans are utterly and asymmetrically dependent upon nonhuman nature—ecological materialism—is fully articulated. This section gives more substance to the central message in the book that dependencies of different scales and types have to be recognized at ontological, epistemological, and ethical levels. Anthropologically, we have to acknowledge and theorize the embeddedness of humankind in natural and social structures through an ecological materialism. In terms of value theory, it means recognizing the existence of values and their complex dependencies among one another as well as their dependence on or independence from humankind. Finally, dependence is expressed in the problem of metascientific stance because worldview clash usually obscures dependence by reading it as interdependence, while science-driven environmentalism systematically obscures the social dependence of knowledge producers. A better metascientific stance allows us to avoid the pitfalls of both of these options and to recognize natural and social dependencies in their multiple forms.

This book is presented as a provocation to environmentalists, philosophers generally, and environmental philosophers in particular. It aims to provoke environmental philosophers to think more deeply about their own conceptions of what environmental ethics and philosophy are and are capable of. Secondly, it aims to provoke philosophers generally to recognize that the field is not restricted to a specific subject matter of concern to just a few scholars in a remote corner of the academic universe. Environmental problems are everyone's problems, and so environmental philosophy is relevant to everyone. Its scope is as wide as philosophy itself, and its content bears on all major areas and problems. Finally, it ought to provoke environmentalists outside of philosophy to become more familiar with the traditional and novel conceptual frameworks and approaches we tacitly rely on in discussing environmental problems, with the hope that they will begin to see the dependence of their own ideas on these frameworks, and will gain the ability to avoid the damaging errors that result from uncritically adopting traditional ones.

PART I

Anthropocentrism and Philosophical Anthropology

CHAPTER ONE

Anthropocentrism, Dualism, and Models of the Human

THE MEANINGS OF ANTHROPOCENTRISM

The introduction described the problem of anthropocentrism as conceived by many environmentalists and located its roots in dualist logic. Many environmental theorists have been aware of the need to expose the limits of traditional conceptions of human nature for decades. Environmental process philosopher Arran Gare argues that "philosophical anthropology is central to ethics and politics" and that an adequate philosophical anthropology "can orient people in their struggle for the liberty to avert a global ecological catastrophe."[1] The ecosocialist Joel Kovel argued that "the notion of human nature is necessary for any in-depth appreciation of the ecological crisis," and ecoliteracy educator David Orr stated that "whatever a sustainable society may be, it must be built on the most realistic view of the human condition possible."[2] In *The Great Work*, ecotheologian Thomas Berry argued that the principle challenge of our time is "to reinvent the human—at the species level, with critical reflection, within the community of life-systems, in a time-developmental context."[3] Radical ecophilosophers as diverse as Arne Naess, Murray Bookchin, and Val Plumwood have also claimed that new conceptions of human being are vital to contesting traditional anthropocentric views and to reinforcing the ecologically informed conception, perception, and evaluation of nature that is called for in environmentalism.[4] This is only a small selection of the many authors who suggest that renewed critical reflection on humankind—or philosophical anthropology—is called for to motivate an effective environmentalist project.

Before engaging more fully with such anthropological conceptions, I want to spend a bit more time disentangling the different connotations often associated with anthropocentrism but underappreciated in the literature, including its broader intellectual-historical context and the argumentative strategies associated with anthropocentrism, in order to show that common responses remain framed by the very dualism they hope to transcend. Doing so makes it easier to see why Plumwood articulates the best critical response to anthropocentrism, and how the anthropology developed in the next chapter builds on her critique with a positive position.

We may identify at least three persistent motifs in discussions of the concept of anthropocentrism. Different definitions of anthropocentrism emphasize the *cosmic, axiological,* and *epistemic* nature of the concept. Fox informally defines anthropocentrism as "the arrogant assumption that we humans are central to the cosmic drama; that, essentially, *the world is made for us*."[5] He approvingly cites the deep ecologist John Seed's claim that "the idea that humans are the crown of creation, the source of all value, the measure of all things, is deeply embedded in our culture and consciousness." These foreground the *cosmological* or ontological dimension of anthropocentrism, and both remarks also invoke a value claim. Other writers better emphasize this evaluative aspect. American deep ecologist Frederic Bender defines *human chauvinism* as "the deeply ingrained assumption that humans have the right to draw down ecospheric integrity—without concern for limits—to satisfy even the most peripheral human desires."[6] This definition locates the problem in *axiological*, evaluative tendencies or errors. Finally, some have claimed that human finitude makes it impossible for humans to relate to the world of nonhumans in anything but a human-centered way, taking anthropocentrism to be "natural and inevitable."[7] These arguments foreground *epistemic* anthropocentrism, the notion that human thinking and experience ineluctably gives a human form to everything thought and experienced. Although we may analytically distinguish the cosmic, axiological, and epistemic themes in these discussions, they are usually entangled in many ways. Take the discourse of the Anthropocene as a current example: it highlights the centrality of humans in a new geological epoch (cosmic), foregrounds the way that humans have prioritized their own interests in development at the expense of the Earth and its creatures (axiological), and usually assumes the apparent inevitability of this species-centeredness (epistemic).[8]

Anthropocentrism critique must be seen as part of the western tradition of self-awareness and self-critique that began long ago in the tradition of

Modern philosophy, although it emerges under a new guise in new historical and cultural conditions of environmental crisis. Although many environmental philosophers are critical of human exceptionalism, few acknowledge that the critique of human exceptionalism is not new to the Modern tradition. For over a millennium, classical European philosophies affirmed the centrality of human beings, in the literal sense that the Earth, and humanity on the Earth, was at the center of the cosmos. The cosmic and axiological conceptions of anthropocentrism are often linked in it. Given this privileged position, in the Modern period it was often seen to be humanity's place to master creation in the fulfillment of its divinely or fatefully ordained role.[9] The rationale of some environmentalist critics was then to show that if cosmic anthropocentrism is false—if human beings are not really at the cosmological center of things—then it follows that human chauvinism in ethics is completely unwarranted. If human beings are not factually at the center of the universe, and if in scientific terms they are not the "pinnacle of creation," then "cosmic anthropocentrism" cannot be used to justify axiological anthropocentrism either. As William Grey put it,

> The intellectual history of the past few centuries can be characterized as pedestal bashing: a succession of successful demolitions of comforting myths through which we have sought to locate ourselves in the world.... First, Copernicus effectively displaced humanity from the *physical* center of the universe. A few centuries later Darwin pointed out that humanity occupied no *biologically* privileged position. Then Freud claimed that one of our fondly cherished distinctive characteristics—*rationality*—was mostly a sham.[10]

Environmentalist critics of anthropocentrism often imply that a further blow to this western self-conception was dealt by the science of ecology, which (following Darwin) treats humans as one organism among others embedded within webs of ecosystemic relations, a treatment that both reveals the harms humans have caused to these systems as well as their complete and utter dependence upon them. Forester and writer Aldo Leopold is one of the best known advocates of this critique, and could be added to Grey's list: the ecology-informed "land ethic changes the role of *homo sapiens* from conquerer of the land-community to plain member and citizen of it," and takes humans to be "only a member of a biotic team."[11] Unfortunately, simply considering new ways of classifying humankind is not enough to motivate environmental social change without a complementary consideration of the moral relevance of the new descriptions.¹² In other words, even when the cosmic centrality of humankind is undermined, anthropocentrists can still identify humanity as a privileged locus of value in different ways. The axiological sense of human centrality usually depends on the evaluation of a few key traits, and is never a matter of simple redescription.

Fortunately, philosophers like Paul Taylor quite clearly revealed the weaknesses in such human exceptionalist accounts early on.¹³ Taylor asked why, if humans are rational and nonhumans nonrational on the traditional western view, this lack of rationality should be taken to justify ill treatment of nonhumans. This would appear to be legitimate only if we have already invested rationality with moral relevance and value, and consider nonrationality a disvalue. Taylor asks what reason we could possibly supply to justify singling out rationality as that human feature that indicates humankind's greater moral worth, and he answers that we are implicitly acknowledging its importance for human life in human societies. "It is intrinsically valuable to humans alone, who value it as an end in itself, and it is instrumentally valuable to those who benefit from it, namely humans."¹⁴ Thus, for Taylor, relative to humans and specific human interests, rationality is an important feature; but taken as one feature of creatures among others in the universe, it has no particular absolute value or disvalue, just as the kangaroo rat can go without water for five years, or the periodic freezing and thawing of the north American wood frog are capacities unique to them and their mode of making a living. In the end, Taylor concludes that claims to human superiority rest on "nothing more than a deep-seated prejudice" and lack justification as morally relevant traits.¹⁵ Thus, Taylor and other writers show that appeals to "cosmic" reclassification alone cannot resolve the problem of anthropocentrism. In other words, along with the question whether such-and-such traits belong to the beings in question, we need to ask just how and why certain traits are valued in the first place. As Plumwood has shown, this is usually accomplished through the work of an a priori dualist logic that has already allotted value to some traits rather than others. Articulating a nonhierarchical conception of difference becomes a major task in response to dualizing thought and practice.

Another type of argument links the cosmic and epistemic dimensions of anthropocentrism. It has often been claimed by nonenvironmentalists (and even some environmentalists) that anthropocentrism is inevitable. This is a peculiarly Modern claim in that it is also rooted in a distinctively Modern metaphysics. The practical challenge this claim raises is that if humans are constitutionally species-centric, then the struggles of environmentalists against the practical effects of anthropocentrism are bound to be fruitless. If humans habitually see, feel, conceptualize, or otherwise experience the world as humans—and are prevented by this from giving the interests of nonhumans or even future generations of humans their due—the conclusion is drawn that human-centered seeing, feeling, experiencing, and by extension, valuing, is inevitable. If we are inevitably anthropocentric, then we cannot help but place human interests front and center and nonhumans outside the moral club. Another variation on the argument holds that values on the whole are subject-relative and human-generated expressions of self-interest or desire. No matter which values are preferred, they are all *my* values, or at least human values, and are, therefore, always relative to me or humanity. The result of this argument is that whatever is valued by humans is valuable to and for humans alone. Thus, here too anthropocentrism seems inevitable.

The argument for the inevitability of anthropocentrism has both empirical and conceptual versions.[16] The conceptual arguments are popular and even more tenacious than their empirical cousins since they involve many unexamined epistemological assumptions. It is important to dwell on this argument right at the start since various forms of it are so widespread in contemporary thinking. These arguments contend that if humans can only experience the world as a human world, we cannot know what the world is like "in itself," or from a nonhuman perspective. (It should be obvious that this argument also counts on the validity of the Modern dualism between humans and nature, where human knowledge is grounded *either* on the human *or* on nature, but not both.) They argue that we are trapped in a prison of human perspective from which there is no escape, whether the walls of this prison are made of limited perceptual faculties, languages, or conceptual schemes. Hence, on this argument, if we can know or value the world at all only through our human categories, this makes it a distinctively human world. If this argument were widely accepted it would be devastating to the environmentalist cause, since in the final analysis it would be impossible to say whether the nature independent of us is sick or healthy. Fox called this the "perspectival fallacy," and Bender the "anthropocentric predicament."[17] A closely related variant in nonenvironmentalist circles calls it *correlationism*.[18] In the environmentalist context, the argument says that if humans cannot experience or think except in a distinctively human way, and if this distinctive way results in nature exploitation, then exploitation or domination of nonhuman nature is unavoidable. Fortunately, the conclusion of this argument simply does not follow because it depends on an equivocation. The first is evident in the use of the term *experience*. The first claim is a tautology, expressing only the *form* of experience. Whatever I think or experience is trivially my thought, my experience, or my value—say, my experience of climate change. But this experience also includes the *content* of experience or thought as well, which is less clearly my own product—did I produce the planet's carbon cycle as such, as well as the disturbance in this cycle that the phrase *climate change* refers to? *What* is thought, experienced, or valued by me is not determined in its substantive features by me, since if this were true, we would never have a mistaken thought and the world would never resist our actions—changing our ideas about climate change would change the climate. Thus, we must maintain a distinction between the image, concept, model, or interpretation of a thing and the thing itself, even if we are not always quite sure where to draw the line between them. The reality of perspective should not be confused with metaphysical idealism. We can create whatever categories we like and try to capture reality with them, but they always model it only partially. Even if "we can't know what the reality of the object in itself is because we can't distinguish between properties which are supposed to belong to the object and properties belonging to the subjective access to the object," it does not at all follow from this that there is no mind-independent reality.[19] Therefore, it is a mistake to infer that we can *only* know our own concepts when we know the world, or to make the even stronger claim that the world is constructed by our categories. Because we often cannot immediately and with certainty distinguish between our contribution and the object's to knowledge, this does not mean that there are no objects existing independently of the mind. This would be to convert an epistemological limitation into an ontological postulate.[20] (I'll discuss this error further in section 3 below.) It should be noted that this common epistemic fallacy also contradicts the principle of asymmetrical dependence outlined in the introduction—the principle of a world not made for us.

On a strongly anthropocentric view, humankind occupies a central place in the cosmic drama, it possesses unique and valuable traits (such as rationality) that no other beings possess, and its action in the world is regulated by its natural and inevitable interests. Addressing these cosmic, axiological, and epistemic aspects of anthropocentrism requires a closer look at dualistic logic, and a broadly applicable analysis of anthropocentrism in light of it. No one has gone further in this area than ecofeminist philosopher Val Plumwood. Understanding the nature and prevalence of this logic will provide us with criteria for evaluating the anthropological views presented by various approaches in environmental philosophy in the remainder of the chapter.

DUALISMS AND THE "LIBERATION MODEL" OF ANTHROPOCENTRISM

As we have seen, the problem of anthropocentrism and the responses to it generally depend upon a dualistic understanding of human and nature. Arguments against cosmic anthropocentrism still contrast human and nonhuman nature in a dualistic, universalizing way. Likewise, arguments around what comes to be called *moral considerability* (what kinds of beings fall under the scope of our moral concern) depend upon a dualistic logic of contrast between a class of beings with certain morally relevant features and one without them (e.g., rationality or nonrationality), which aims either to assimilate the other in order to legitimate instrumentalization, or to honor it by excluding it from the class of exploited things. Similarly, arguments for the inevitability of anthropocentrism depend on the sharp distinction between objectivity and subjectivity also construed in a dualistic manner (i.e., either the environment exists independently of the mind or we construct it). Each of these problematics is characterized by dualist logic. Revisiting Plumwood's analysis in more detail below is necessary to extract the criteria that a more productive environmentalist philosophical anthropology, value theory, and ontology must satisfy, and we will then be equipped to measure existing views against them.

While many environmental philosophers' responses to anthropocentrism are important, they are often limited in the sense that they do not fully reveal the way in which the different cosmic, moral, and epistemic facets of anthropocentrism are systematically interconnected. Plumwood's does this, and her "liberation model of anthropocentrism" is based on her account of the pervasive dualistic logic shaping western conceptual frameworks. She defines and explains this logic in several different works, and I refer the reader to these richer sources for a more exhaustive account.[21] Here I provide only an outline. What makes hers a liberation model is the claim that anthropocentrism must be seen in parallel with the critiques of Eurocentrism, ethnocentrism, androcentrism, and other "hegemonic centrisms," because they are all derived from the same dualistic logic of domination. It has long been recognized that most of Western thought rests on sets of opposed categories, and Plumwood gives an account of the way that these oppositions are philosophically and popularly conceived in order to support arguments justifying discriminatory judgments and behavior, whether in terms of gender, race, class, or species (among others). Anthropocentrism critique understood in terms of this logic can be conceived as the liberation of more-than-human nature from social domination in the same way that feminism seeks women's liberation from oppression, for example. This is possible because "dualisms are not universal features of human thought, but conceptual responses to and foundations for social domination."[22] Plumwood ingeniously allies social and environmental struggles by identifying their common enemy, providing a robust theoretical framework for a whole range of social movements.

As noted in the introduction, the five features of the logic of domination are (1) radical exclusion or hyperseparation, (2) homogenization or stereotyping, (3) denial or backgrounding, (4) incorporation or assimilation, and (5) instrumentalization. The dualist logic takes advantage of existing categorial dichotomies and renders them in an oppositional way. Dualized pairs include, but are not limited to, the following:

Culture / nature	Freedom / necessity
Human / animal	Universal / particular
Male / female	Human / nature (nonhuman)
Mind / body	Civilized / primitive (nature)
Master / slave	Production / reproduction
Reason / matter (physicality)	Public / private
Rationality / animality	Reason / emotion
Mind (spirit) / nature	Soul / body
Self / other	Subject / object

Each of the terms on the left in each column is regarded as "superior" to each of those on the right. The list is incomplete, but forms the basis for major forms of oppression in the west. "The dualisms of male/female, mental/manual (mind/body), civilized/primitive, human/nature correspond directly to and naturalize gender, class, race and nature oppressions respectively, although a number of others are indirectly involved."[23] They are made mutually reinforcing largely through analogies that create equivalences or map the pairs onto others. For example, "the assumption that the sphere of the human coincides with that of intellect or mentality maps the mind/body pair on to the human/nature pair, and, via transitivity, the human/nature pair on to the male/female pair."[24] Such linkages insure that the terms female-nature-body-primitive are repeatedly associated, and are contrasted with male-human-mind-civilized, the privileged terms. The development of these contrasts has been a culturally specific, historical process. According to Plumwood, the west has become especially anthropocentric as a result of this dualist logic.

(1) *Radical exclusion* is both a matter of categorization and evaluation, involving "not just difference, but defining the dominant identity against or in opposition to the subordinated identity, by exclusion of [its] real or supposed qualities," thereby justifying domination. She continues:

> Separate "natures" explain, justify, and naturalize widely different privileges and fates between men and women, colonizer and colonized, justify assigning the Other inferior access to cultural goods, and block identification, sympathy, and tendencies to question. A sharp boundary and maximum separation of identity enable the beneficiaries of these arrangements to both justify and reassure themselves.[25]

We are dealing not simply with "difference," but with a hierarchical and invidious conception of difference. In the case of the dualistically construed categories human and nature, properties like mind, rationality, purposive agency, and moral capacity are attributed to humans and denied to nonhuman animals, and identification or sympathy with these Earth others is blocked.[26] Anthropocentrists emphasize the nonrationality of the nonhuman in order to justify unrestricted domination. The appropriate response to such discriminatory categorization involves "reexamining and reconceptualizing the concept of the human, and also the concept of the contrasting class of nature."[27] The first criterion for an appropriate philosophical anthropology for environmentalism is thus a *nonhierarchical* and nonseparatist conception of human and nonhuman difference.

(2) *Homogenization* or *stereotyping* minimizes or eliminates differences within groups, rendering men universally different from women, or humans from nature, for instance. "Thus essential female nature is uniform and unalterable. The colonized are stereotyped as 'all the same' in their deficiency, and their social, cultural, and religious and personal diversity is discounted."[28] In anthropocentrism, all nonhuman natural beings are conceived as alike, lacking mind, consciousness, or sentience, and the diversity of nature is obscured. Interchangeable and replaceable physical units, resources, or services compose the nature available for appropriation by humans, and underestimating both diversity and complexity in nature is implicit, for example, in mechanistic approaches that rely on the metaphysical dualism "consciousness and clockwork." The dominant group is also homogenized in symmetry with the otherized group. This pattern is another condition

of the universalizing environmentalism claims that the homogenous class of all humans is responsible for environmental problems. The second criterion for an adequate anthropology is thus the *dehomogenization* of the classes *human* and *nonhuman*, attending to diversity at all relevant levels of analysis.

(3) To explain *denial of dependence* or *backgrounding*, Plumwood notes that "once the Other is marked in these ways as part of a radically separate and inferior group, there is a strong motivation ... to represent them as inessential. Thus the center's dependency on the Other cannot be acknowledged."[29] We cannot overstate the importance of this aspect of dualizing logic in relation to denial of human dependence on nature. "Denial often is accomplished via a perceptual politics of what is worth noticing, of what can be acknowledged, foregrounded, and rewarded as 'achievement' and what is relegated to the background."[30] The contributions of nonhuman nature and the colonized "are denied as the unconsidered background to 'civilization'" and "they are represented as inessential." For anthropocentrists, nature is rendered inessential, and its own needs do not factor into the judgments and actions of humans, and do not impose limits on human activity. Plumwood herself repeatedly discussed the significance of anthropocentric denial of dependence: "The denial of dependency combines with the western master story of human hyperseparation to promote the illusion of the authentically human as outside nature, invulnerable to its woes."[31] As she explains, the "authentically" human is what is maximally different from the contrasting term, woman or nature or animal, such that the essential human feature is precisely what is *not shared* with the Other. But this dualistic, disembodied, and disembedded conception of the authentically human "encourages a massive denial of dependency, fostering the illusion of nature as inessential and leaving out of account its irreplaceability, non-exchangeability and limits."[32] When an entire culture is able to regard its reproduction and maintenance as *independent* of natural constraints, we are dealing with an anthropocentric culture that has designed its social institutions to systematically neglect ecosystemic dependencies, as well as dependencies on otherized humans who support and maintain the system. "Dependency on nature is denied, systematically, so that nature's order, resistance and survival requirements are not perceived as imposing a limit on human goals or enterprises."[33] The third criterion thus must include acknowledgment of the basic principle of dependence. Since this entails

realism or naturalism, I'll show that it also limits the range of acceptable theories of human nature.

(4) *Incorporation* or *assimilation* refers to the definition of the Other in terms of lack. Plumwood refers to the pioneering work of Simone de Beauvoir: "Humanity is male and man defines woman not in herself but as relative to him; she is not regarded as an autonomous being.... She is defined and differentiated with reference to man and not he with reference to her; she is the incidental, the inessential as opposed to the essential."[34] Likewise, nature is defined in terms of its lack of human qualities, rather than in terms of its own positive attributes. This is another aspect that a nonhierarchical concept of difference would address. On its own, nature may also be conceived as disorderly, raw, requiring human-aided "development" to reach its full potential, encouraging projects of appropriative intervention. Concepts of the human as steward or as perfecting nature imply such a conception of nonhuman nature. The fourth criterion for responding to dualisms is a corollary of the first, which insists that whatever is considered the contrast class of humankind must be regarded in terms of its positive properties, possessing its own human-independent characteristics and forms of agency.

(5) Finally, *instrumentalization*, treating the Other as nothing but a means to the center's ends, may be seen as both the result and objective of dualizing logic. If the other is stripped of agency, values, purposes, or intentions of its own, the way is open to impose another's ends on it. "Nature's agency and independence of ends are denied, subsumed in, or made to coincide with those of the human, the source of all value in the world."[35] Since it is empty of agency, and hence purpose, humans are free to impose their own purposes on nonhuman nature. Other-than-human nature becomes an instrument, a collection of means with which to achieve human ends. According to a recent definition, *ecosystem services* are "the benefits humans derive from ecosystems," without reference to the benefits that nonhuman nature itself also receives from the continued good functioning of those services. It is important to note in relation to ontologies of nature that while endowing nature with minimal agency may be a step forward, taken in isolation from patterns of valuation it cannot resolve the issue of exploitation. Even when Others (e.g., other humans) are attributed full agency, the ends or goals of the "superior" class are easily imposed on them, cancelling or denying their primary self-determination. The fifth criterion for a critical environmental anthropology is that it cannot regard instrumentalization

of nonhumans (or even of other humans) as an inevitable feature of the human condition.

The liberation model is far more relevant to environmentalism than classical universalizing models because it may be directly conjugated with political and social struggles, forming a substantive conceptual framework for political ecology. As Plumwood notes, "Since liberation models do have a practical and behavioral orientation (after all, they have emerged mostly from the reflection of people involved in active social change in liberation movements), the model can close the gap between ecophilosophy and ecopolitics to validate and illuminate theoretically the sorts of practical responses adopted in green activism."[36] Secondly, it allows us to see the limitations of the previous models. All five characteristics of the dualist logic have to be addressed if this logic is to be dismantled, since they are interrelated in a comprehensive structure of thought and action in everyday human experience. In other words, where we find one, we usually find the others. Given the tenacity and implicit nature of these interlocking elements, however, isolating one aspect of the logic does not automatically dismantle the overall pattern. Conventional environmentalist responses have emphasized different facets of the logic, but none have addressed them all. Many approaches, including deep ecology, for example, take as their leading problem the hyperseparation of human and nature, and strive to develop an ontology and concept of the human which overcome this separation by speaking of the identity or unity of human and nature. The "self-realization" of this unity on the part of the individual is taken to be the cure for all aspects of the problem, including denial of dependence, assimilation, homogenization, and instrumentalization. But the resolution of these problems does not follow unproblematically from the simple assertion of unity.[37] How do we depend on something we are identical with? Does nonexploitative behavior really follow directly from metaphysical ideas about unity? Other approaches may begin from a naturalistic perspective, where hyperseparation has been addressed by arguing that humans are evolutionarily continuous with nonhuman life on Earth. The evolutionary narrative is meant to show that humans are products of the same natural principles that produced all other forms of life on Earth. However, while centrally important for an ecological conception of humanity, not only is pointing out similarity of causal origin an utterly ineffective strategy to convince people to care for the Earth, it too leaves out the four other dimensions. It is just as important, if not more important, to emphasize the degree to which human beings complexly depend upon beings different from and similar to themselves, regardless of origin, for their continued existence in reproductive, ecological time. Evolutionary narratives about how we have the same genes as, and so are continuous with, other organisms, while generally helpful to overcome some sense of ontological separateness, are usually not at all helpful for foregrounding the degree to which we depend on other organisms and the integrity of ecosystems for our lives. Therefore, approaches that insist exclusively on continuity or substantive identity (aiming to resolve hyperseparation alone), whether metaphysical or naturalistic, are limited in what they can do for environmental philosophy. The more features of dualism a given approach tackles, the better. However, in all cases where fewer than all five are addressed, I doubt whether their proponents have fully appreciated the depth, scope, and consequences of dualism for our tradition. Therefore, those positions are regarded as stronger that recognize and address each of these aspects.

In the first two chapters I outline a conception of the human that responds to each of these features of dualistic logic. Environmental philosophers, including Plumwood, have often hinted at alternative theories of human nature that would better enable their work but did not work out these theories themselves. In the remainder of this chapter, I explore a few conceptions of human nature implicit or explicit in environmental philosophies, and criticize those that diverge from the liberation model in the direction of idealism, as well as in the direction of reductionist naturalism. In the next chapter I sketch a naturalistic philosophical anthropology that gives substantive meaning to the claims of dependence, situatedness, and embeddedness in a larger nonhuman material world.

MODELS OF THE HUMAN AND ENVIRONMENTAL KANTIANISM

In a recent book, philosopher Rosi Braidotti identifies the "posthuman condition" as one in which "the basic unit of common reference for our species, our polity and our relationship to the other inhabitants of this planet" can no longer be defined in terms of classical European humanism, and she includes environmentalism in a cluster of other posthuman, often anti-humanist, discourses. She cites the explosion of work in animal studies, ecocriticism, environmental humanities, science and technology studies, human enhancement, cognitive sciences and information technologies, and finally the continuing development of critical philosophies of the subject, as evidence of the posthuman condition.[38] What Braidotti calls "posthuman theory" might also be called—employing a more traditional term—philosophical anthropology.

Braidotti notes that posthuman studies have in common a critique of the classical European image of rational Man, able to transcend his own specificity in order to speak in the name of all humanity. European humanism is based on the confidence that by means of reason and cultural (scientific) development, humans can individually and collectively perfect themselves and expand their capabilities to achieve unbounded progress.[39] She argues that the anti-humanism of much 1970s European theory is meant to disengage the conception of human being from this universalism and progressivism. Feminism, post-structuralism, postcolonialism, and critical race theories do this by refusing hierarchical conceptions of difference, by recovering "situatedness," cultivating sensitivity to power relations, and formulating a positive concept of difference. As we have seen, environmentalists have done this by insisting on human continuity with and dependence on, rather than superiority to and separation from, nonhuman nature.

Braidotti is obviously not the first to engage in this kind of wide-ranging philosophical anthropology. German philosophical anthropologist Max Scheler (1874–1928) and other anthropological writers in the early twentieth century can be resources for environmentalists since they too attempted to sidestep the dualistic divisions between body and mind, nature and culture, and even the division between the natural and human sciences. Against the cultural backdrop of anxious social and political uncertainty—which we could argue is also characteristic of this era of globalization—they sought to redefine the human in order to reorient individual and collective struggles.[40] In doing so, they sought to avoid reductionistic scientific views as well as extreme human exceptionalist ones. In the essay "Man and History," Scheler created a typology of models of human being that in 1926 he thought were still effective in many different discourses, and in contrast to which he formulated his own philosophical anthropology.[41] It is remarkable how well the current models within cultural discourse generally (including academic discourse), and environmental discourse in particular, exemplify the types already delineated and assessed by Scheler. In the following table I present a taxonomy mostly based on Scheler's own.

A few views from the environmentalist literature falling into these taxonomic categories will be assessed in terms of whether they satisfy the criteria of the liberation model of anthropocentrism, including a nonseparatist and nonhierarchical conception of difference and an affirmation of asymmetrical dependence (in place of radical exclusion and denial of dependence). This typology includes six models of human being, and a seventh will be added in the next chapter. A brief summary of the key features of each, with some sense of their environmental implications, will have to suffice.

Exploring the problems with these common views is highly instructive for avoiding the pitfalls of typical responses to dualism. These models provide a framework for a critique of philosophical anthropology that moves in broad strokes from an ecologically problematic picture of humans as disembodied souls, reasoners, or perfecters of nature, toward one that is naturalistic, nonteleological, and nonreductive. In what remains here I'll briefly discuss two common options: rationalist (Greek) and pessimistic naturalist (Panromantic) positions. The rationalist position subtly preserves the dualist hierarchy intact, while the Panromantic position simply aims to invert it. Neither avoids it.

While still opposed to anthropocentrism, environmental rationalism tends to emphasize *reason, rationality, consciousness, or deliberate choice as the defining feature of human nature.* For these approaches it is our rationality that makes us different in kind from other living things, a difference that endows us with the capability of effecting real change in the world. Kantian environmental ethicist Paul Taylor typifies the position: "Nothing prevents us from exercising our powers of autonomy and rationality in bringing the world as it is gradually closer to the world as it ought to be."[42] Taylor's view is basically Kantian—despite his embrace of an evolutionary model of humankind—and Kant's view is unabashedly dualistic. As a result, Taylor's conception of the relation between nature and reason in the human being is at best equivocal. Although he begins with the ecological idea of human evolutionary continuity with all life in order to motivate his ethics, its implications for ethics are not at all direct.

> We share with other species a common relationship to the Earth. In accepting the biocentric outlook, we take the fact of our being an animal species to be a fundamental feature of our existence. We consider it an essential aspect of "the human condition. . . ." The laws of genetics, of natural selection, and of adaptation apply equally to all of us as biological creatures. In this light we consider ourselves as one with them, not set apart from them.[43]

As mentioned above, one of the implications of this awareness includes fundamentally rejecting human chauvinist understandings of our place in the world as groundless and as inconsistent with the biocentric worldview. He emphasizes evolutionary continuity over difference here, and in light of a long tradition of radically separating humanity from nature, he is justified in doing so. While he does not altogether deny "our special abilities or our uniqueness," he insists

Table 1.1. Models of the Human

Model	Outline	Environmental Implications
Judeo-Christian	Ensouled body, fallen into sin with the hope of redemption, created "in the image of God," placed between beast and angel, having dominion over the creatures of the Earth.	The image attacked early on by the historian Lynn White Jr., who claimed that the theological "dominion thesis" sanctioned and empowered continued human exploitation of nature.[i] Denies significant dependence of the human on the other-than-human material world.
Greek	Distinctive capacities *nous* (mind) and *logos* (language) make humans the noble "rational animal," raising them above the brutes and making the divinities their kin. On this view, reason or spirit is considered timeless, universal, and endowed with the power to effect changes in the world.	Fuels most anthropocentric arguments, since it insists on the special ontological status of rationality in the cosmos, creating the radical exclusion between reason and its "others" (body, emotion, particularity, animal, woman, nature, etc.) on which so many dualisms are built.
Naturalistic	Associated with positivism, pragmatism, and the sciences, holds that human beings are one evolved type of animal among others and ontologically on a par with them. Human capabilities are more sophisticated versions of capacities which can be found among other forms of life, only "different in degree" of complexity.	Evolutionary narrative levels the ontological playing field and includes the human species in the contingent history of organic development on Earth. Often used to overcome the radical exclusion and human chauvinism to which the Greek and Judeo-Christian views usually lead. Reductionist and nonreductionist options exist.
Panromantic	Naturalistic, but regards central human capacities that European humanism claimed led humans to the triumphant development of high culture and civilization as dangerous and debilitating rather than noble and praiseworthy.[ii]	Primitivist environmental writers blame the environmental crisis on "civilization," indicting development of agriculture, phonetic writing, organized religions, etc., and advocate a return to earlier cultural lifeways as solution to the ecological crisis.
Teleological	Metaphysically (as idealism or materialism) gives humanity a prominent place and role in the cosmos; humanity is the voice, mind, expression, culmination, steward, perfecter, or representative of "nature as a whole."	"Humanity is nature become self-conscious" overcomes radical exclusion and supplies some normative force. Variations of this view are surprisingly widespread among radical environmentalists.

Table 1.1 cont'd

Model	Outline	Environmental Implications
Postmodern	While postmoderns typically deny that there can be such a thing as "human nature,"[iii] one is nevertheless everywhere implied in "philosophies of the subject" whose image of the human is of a fragmented, situated, embodied, gendered, subjectified, disunified being strewn across the nodes of linguistic, material, economic, political, and symbolic networks.	Despite their attention to conditions of subjectivity, they continue to background and deny human dependence on an independently existing nonhuman nature.[iv]

i See Lynn White Jr., "The Historical Roots of our Ecologic Crisis," *Science* 155, no. 3767 (1967): 1203–1207.
ii Representative figures cited by Scheler included Ludwig Klages, Friedrich Nietzsche, and Sigmund Freud. Nietzsche and Freud are included because they argue that "civilization" is the effect of *repression* of life drives or unconscious forces, rather than the product of an independent faculty like "reason." It is generally a philosophically pessimistic stance.
iii According to John Protevi, "For many years, a large part of the left adopted social constructivism to fight the good fight against racist and sexist constructions of human nature. But they threw the baby out with the bathwater by banning any discussion of human nature. We cannot escape a new serious and important discussion on human nature." (*Political Affect: Connecting the Social and the Somatic* [Minneapolis: University of Minnesota Press, 2009], 188).
iv There are many variations on this general model of subject as product of structural factors. The negative postmodern image obviously has no prospect of forming a viable liberation view for environmentalism. See Braidotti (op. cit.) for more discussion of the variations. For some discussion of the tensions between postmodernism and environmentalism generally, see Kate Soper's *What is Nature?* (Cambridge, MA: Blackwell, 1995), as well as Arran Gare's *Postmodernism and the Environmental Crisis* (London: Routledge, 1995).

that "this biological aspect of our human existence places certain *requirements and constraints on the manner in which we conduct ourselves* in relation to the Earth's physical environment and its living inhabitants."[44] The question is what these "requirements and constraints" are that our biological aspect seems to impose on us. In this context, he refers to Leopoldian "nature-as-teacher" ideas as well as sociobiology, both of which, he claims, illegitimately attempt to derive ethical norms from natural facts and regularities. He ultimately admits that "understanding ourselves as biological entities, however, does not provide us with any particular directives as to how we should conduct our lives. Our proper role as moral agents is not deducible from facts about our biological nature."[45] While the biological aspect of human being and its moral relevance are affirmed, the relation between the reflective, rational biocentric outlook and action simply remains indeterminate. The question is whether "our biological nature [is] at all relevant to the choices we must make as moral agents, and, if it is, in what ways is it relevant?"[46] In the end, although human beings are both biological creatures and rational persons, only as persons do we give ourselves the moral law. "A person's autonomous choice places a (conceptual or logical) boundary on what can be explained in *human* life by natural selection and adaptation."[47]

This is a thoroughly Kantian conception. Persons are centers "of autonomous choice and valuation. Persons are beings that give directions to their lives on the basis of their own values."[48] This includes choice of short- and long-term goals in light of interests and purposes, choice of what ends to seek, and the power to rank order them, judging them relative to the person's overall well-being. "Full personhood" is "the most complete actualization of the powers of autonomy and rationality required for choosing one's own value-system and for directing one's life on the basis of that value-system."[49] Personhood comes in degrees (from youth to adult), can be lost and regained (through, e.g., cognitive dysfunction), and is ostensibly an achievement (not a given) of the human organism under the right conditions. Unfortunately, while Taylor explicitly argues against human chauvinism and human separatist conceptions, he nevertheless preserves a Modernist conception of the human agent by accepting this basically Kantian distinction between human biology and human personhood. While he does not exclude the biological from his account, its status is rendered highly ambiguous since it is not directly central to our conception of what is uniquely human. In the end, if persons are fully autonomous, it is we who decide whether or not nature is valuable, and our biology is ultimately irrelevant for moral determinations, since this would result in a heteronomous ethics (to use Kant's term). His concept of the human being includes a distinction between the biological and rational, but does not provide any satisfactory way of understanding how practical reason is conditioned by or asymmetrically dependent upon its own biological nature or on its environment.

As Plumwood astutely noted, for many ethicists

> the mental rather than the biological have been taken to be characteristic of the human and to give what is "fully and authentically" human. The term "human" is, of course, not merely descriptive here but very much an evaluative term setting out an ideal: it is what is essential or worthwhile in the human that excludes the natural. It is not necessarily denied that humans

have some material or animal component—rather, it is seen in this framework as alien or inessential to them, not part of their fully or truly human nature. The human essence is often seen as lying in maximizing control over the natural sphere (both within and without) and in qualities such as rationality, freedom, and transcendence of the material sphere.[50]

This may be a bit unfair to Taylor, but the tension in his position is palpable: he has claimed both that humans are "one with" nonhuman nature, and yet that rational personhood is what is most characteristically human. Another manifestation of dualism in his account is the reason-emotion dichotomy, which aligns reason with universality and emotion with particularity, and which conceives of ethics as a matter of rule following and universal obligations for autonomous individuals rather than a matter of empathy, care, preserving relationships, or prioritizing some values over others in situated social contexts.[51] (I'll say more about this in part 2.) Despite the fact that Taylor aims to develop a biocentric ethic, the ethic is still based on more subtle forms of the traditional dualisms.[52]

Twentieth-century European philosophers, of course, also spent a lot of time trying to undermine prominent dualisms such as these. Neil Evernden's book *The Natural Alien* was one of the first attempts from the Continental philosophical camp to come to terms with the environmental crisis. In it, he addresses anthropocentrism primarily through the lens of phenomenological authors Martin Heidegger and Maurice Merleau-Ponty. Environmental ethics per se does not form part of this program, and it is considered an alternative to mainstream environmental ethics like Taylor's. Changes in behavior would result not from a new normative ethical theory alone, but from the adoption of a novel ecological worldview. As with Taylor, however, lingering unthematized dualisms continue to structure the account.

Theories like this are nuanced but may still be classified as rationalist because of the power they allot to human consciousness in determining what "nature" is. Evernden explicitly states that nature "exists as a structure of meaningful distinctions," and his book is "about those 'meaningful distinctions' that give us an environment."[53] It is expressly not about nonhuman nature as it is "in itself," then, but only about human ideas of nature. Evernden claims that "the source of the environmental crisis lies not without but within, not in industrial effluent but in assumptions so casually held as to be virtually invisible."[54] By adopting a phenomenological approach, Evernden thinks, the Cartesian dualist and mechanist worldview is allegedly undermined. Phenomenological method, developed by Edmund Husserl (1859–1938) as a tool for the unbiased description of the full field of human experience, ostensibly closes the epistemological gap between consciousness and thing, subject and object, by insisting on the "intentional" nature of consciousness. According to this approach, consciousness effectively constitutes its "objects." Evernden cites an example from fellow ecophenomenologist Erazim Kohak to illustrate the notion of intentionality. A smoker in need of flicking off her ash casts about and finds no ashtray, but with a little more effort, finds a seashell and knocks off her ash, for "[sh]e did not 'find' an ashtray 'in the objective world'; there was none there to be found. Rather, [s]he constituted an ashtray in [her] act."[55] Such an object "can only *be* if there is a smoker who has a world."[56] This phenomenological point about intentionality is relevant to environmentalism because a worldview is the comprehensive horizon within which a mechanistic objective world or, in contrast, a community of subjects is constituted. "If we could deprive the smoker of [her] ability to conceive of ashtrays as effectively as we have removed our collective ability to constitute a world of subjects, the population of ashtrays would be similarly decimated. In effect, what cannot be conceived cannot exist—for us."[57] The upshot is that because we are (and things are) how we conceive of ourselves (and them), and we conceive the world and ourselves *as* pieces of mechanism rather than *as* subjects or agents, a world of agents does not exist. "The eye altering alters all."[58] But since we have the power to choose our worldview, we should choose to view the world as full of agencies. Since "consciousness encloses its object" and "the world is an ingredient of consciousness," all meaningful distinctions arise from consciousness or human-in-the-world.[59]

As I mentioned in section one, one major flaw of this type of argument is that it relies on an equivocation in key terms. In this case, by defining consciousness as "consciousness of" something, Evernden conflates the ashtray as ideal meaning (intentional object) and ashtray as real entity. The meaning *ashtray* may have been constituted by the smoker in the act of ashing into the shell, but the shell itself as *transobjective* (as more than a mere "object for us") remains, whether or not any sensitive being is there to interact with it. It may only "be" an ashtray if a smoker is in its vicinity, but it already "exists," beyond the image had or meaning given it by an observer. The point can be generalized to more popular social constructionist versions of the argument. As Plumwood herself noted, "meanings and concepts may be cultural products, but it does not follow that what they designate are also, or we are forced to the extreme idealist conclusion that the entire universe, including distant stars we know nothing about, is a cultural construct."[60] Evernden denies that such a distinction can even be made, and this denial is a serious weakness for environmentalists who must acknowledge the principle of dependence.[61] While we should appreciate the well-intentioned encouragement to see the world as populated with nonhuman subjects or agencies in contrast to dead mechanisms to be exploited, we have to recognize that this too is a basically Kantian approach that distributes determinative power to constitute nature to the human pole alone. Asymmetrical dependence is at the very least obscured, if not denied outright. Overcoming this subjective idealist or correlationist conceit is also necessary for developing a critical environmental philosophy.[62]

To briefly review, the forms of environmental rationalism and idealism sketched here fail to satisfy the criteria laid out in Plumwood's liberation model of anthropocentrism. While they are different in important ways—Taylor's rationalism and Kantian ethics is put into question by phenomenological trends in twentieth century philosophy—they are also similar in their continued adherence to a conception of reason or mind as a special determining power governing or constituting reality. In this way, they fail to acknowledge the environmentalist principle of dependence, that is, humankind's asymmetrical dependence on the more-than-human. Thus, the extreme irony of the idealist conception of worldview, meaning, and ontology in an environmentalist context is that it is no less anthropocentric than any of the traditional views (and might in fact be more anthropocentric in some ways).[63] This general philosophical trend has become so commonplace that it is challenging to recognize it as the radical anthropocentric idealism that it is. It says that the world does not exist without us, and that it is made by us, and to a large extent, for us. The intellectual inconsistency of this environmentalism is akin to the practical inconsistency of the environmentalist who drives an electric car whose electricity is primarily produced by coal. It is both a substantive (performative) and logical contradiction. Instead of this patent anthropocentrism, I argue that in order to develop a critical environmental movement and philosophy, we must first acknowledge that we occupy a world not made for us. The ultimate resolution of the environmental crisis cannot be to constitute nature differently. The point is to acknowledge that we do not constitute it in many of its most important modes of existence. Our bodies and minds depend on the more-than-human to provide the content and even forms of thought, while it does not depend on us to be itself. Every epistemology that grants to the subject the ontological power to determine that which is denies this basic asymmetrical relation. When environmentalists talk about how the

Earth and its creatures do not need us, but we need it and them, they are expressing this dependence in a crystal clear fashion.

PLEISTOCENE DREAMS

We might then ask whether explicitly naturalistic approaches do any better at avoiding dualism and acknowledging dependence. Because environmentalism is a popular social as well as intellectual movement, critical environmentalists have to engage as much with popular ideas as with abstruse argumentation since they play a significant role in forming the ecological imaginary of large environmentalist constituencies. These also entail specific anthropological views that must be carefully examined in order to avoid the errors of dualism and denial of dependence if we hope to generate a better philosophical anthropology for environmentalism.

Among the more radical environmentalist views, some form of primitivism has remained popular since the 1970s. Here primitivism is considered a Panromantic model, a pessimistic naturalistic approach to humankind that *regards the core human capacities that European humanism claimed led humans to the triumphant development of high culture and civilization to be dangerous and debilitating rather than noble and praiseworthy*. Within the broad framework of anthropocentrism critique, primitivism identifies *reason* and *civilization* as characteristics that have made humankind a scourge on the Earth. But these Modern, destructive forms of reason and civilization have not always existed. Therefore, they argue, we should look at that time in human history where they did not yet exist as such. This is identified as the Pleistocene, the period between around 2.5 million and 10,000 years ago during which many different human and humanlike species existed and evolved, including *Homo erectus* and *Homo sapiens*. If there is evidence that these humans (and their Modern foraging-hunting descendants) lived a less environmentally destructive lifestyle, there may be features of it that are worth imitating or recovering in the contemporary world. At face value, the argument does not seem entirely implausible. There is some evidence that foraging-hunting peoples are less ecologically destructive than their "civilized" counterparts, although many decades of historical and anthropological research have certainly taught us not to romanticize indigenous harmony with nature. The arguments of the primitivists nevertheless oppose gathering-hunting cultures to "civilization" in a dualistically coded way in an attempt to overturn the Enlightenment opposition between savage and civilized, such that gathering-hunting lifeways will be regarded as superior and Modern civilized life as inferior. This kind of inversion is a typical strategy of Romanticism. Such reactive naturalisms generally aim to deflate or dethrone reason, culture, or civilization, and substitute for them intuition, natural impulses, and indigenous ways of life. More subtle thinkers, such as ecologist Paul Shepard (1925–96), attempt to overcome rather than simply reverse the persistent dualism, though, as I argue, even these attempts are unsuccessful. Here I'll take Shepard's views as representative.[64] Since Shepard explicitly relies on arguments in the literature of sociobiology and evolutionary psychology for support, I also briefly examine forms of naturalism about human being that stem from them. These views will be very instructive for outlining a nonreductive naturalist anthropology in the next chapter. We will see that these views have an initial naturalistic appeal as a response to human hyperseparation from nonhuman nature, but that the assumptions on which their naturalism is based are highly questionable.

Opposing those who sang in praise of reason and the unlimited progress of civilization, Shepard called ecology the "subversive science" because he thought that it clearly demonstrated that human civilization had gone terribly wrong. The title of his last book, *Coming Home to the Pleistocene*, effectively summarizes Shepard's basic intuition as well as his theoretical point of departure:

> In the face of predominant anthropocentric values, the vision of natural humankind seems eccentric, regressive, even perverse. Our idea of ourselves embedded in the context of the shibboleth of growth places us at odds with the notion of kinship with nature. When we grasp fully that the best expressions of our humanity were not invented by civilization but by cultures that preceded it, that the natural world is not only a set of constraints but of contexts within which we can more fully realize our dreams, we will be on the way to a long overdue reconciliation between opposites that are of our own making.[65]

For him, "the vision of natural humankind" means specifically that human brains and behaviors are best seen as well adapted to living a low-impact existence in an open savanna, Pleistocene environment, in which human beings spent the longest period of their evolutionary history. Civilization, which on the primitivist definition includes institutions like sedentary agriculture, phonetic writing, mercantilism, measured time and space, and hierarchical social organization, amounts to a state of alienation from nature, a destructive form of domestication bent on eradicating the Stone Age "savage" in us. "Coming home" to the Pleistocene is seen as a way of rejuvenating latent human spontaneity and health, as well as of relieving nonhuman others of domination and exploitation. Looking as best we can into "the unique mind of our hunter/gatherer ancestors" has a redemptive function in Shepard's work.[66] The aim is "to live wholly an authentic Pleistocene existence," which would be "that final recovery of our truest being."[67] His argument is that "the greater the degree to which a person or society conforms to [the example of] our Paleolithic progenitors and their environmental context the healthier she, he, they, and it will be."[68] Shepard lists a series of traits and institutions that dualistically pit the Stone Age against civilization, such as social egalitarianism, a gift economy and spirituality, little or no heritable social rank, small community sizes, decentralized power, and formal recognition of stages in the life cycle for gatherer-hunters. Civilization exhibits the strong need to exert control over nature, domestication of wild forms, stress caused by living in larger groups and the threat of food scarcity, the rise of hierarchical and theocratic states, loss of autonomy for the majority, centralized power, domination of nature, and transcendent spiritualities showing contempt for the Earth.[69] While I cannot do justice to the care and depth with which these contrasts are treated by Shepard, we should at least reflect on the assumptions on which the approach is based.

As a preliminary methodological point, anthropologists have long warned against using the method of empirical generalization over many diverse cultures. To the contemporary historian or anthropologist there is as little "civilization in general" as there is a "savage mind" in general. The contrasting characterization of both "civilization" and "the unique mind of our hunter/gatherer ancestors" might be seen as a rationalistic, dualistic abstraction. It is a universalization of both human nature and culture to the detriment of particularity and attention to situatedness indispensable for resolving concrete social and environmental conflicts. Although it seems warranted if lessons are to be learned from such cultures, this method must be handled with care.[70] Secondly, this generalizing approach entails a questionable conception of history as a grab bag of modular cultural elements. Culture is regarded as a mosaic system where "each trait is portable yet embedded, constituting with other bits a whole, a complex, that can be disarticulated and reassembled." Given this mosaic model of culture, it may seem possible that we (civilized, unhealthy westerners) are able to return to a Pleistocene lifestyle, since "elements in those cultures can be recovered or re-created because they fit the heritage and predilection of the human genome everywhere, a genome tracing back to a common ancestor that Anglos share with Hopis and Bushmen and all the rest of *Homo sapiens*. The social, ecological, and ideological characteristics natural to our humanity are to be found in the lives of foragers."[71] With reference to the dualistic pairs we have been dealing with, we see that methodological generalization leads to a number of features characterizing foraging groups, and then these shared features are coded *natural* while others are coded *cultural*. Culture, however, becomes more and more insubstantial throughout the account, an echo of the deep bass undertone of nature. "We can go back to nature, as I wrote in 1973, because we never left it.... We cannot avoid the inherent and essential demands of an ancient, repetitive pattern as surely as human embryology follows a design derived from an ancestral fish."[72] This attitude toward appropriating the past depends on a questionable way of thinking about the priority of nature to culture, effectively overturning the Modernist privileging of reason over nature. How this priority is conceived is one of the most important issues for naturalistic environmentalism. The Romantic opposition between nature and civilization, giving nature primacy, must also carefully avoid biological determinism, or it makes the choice of harmful or maladaptive lifeways—and our ability to choose better ones—puzzling. Deterministic conceptions of nature (if consistently held) are always unsuitable for effective projects of social engagement because they are usually in tension with our ability to choose. If humankind is hardwired (to use a popular and highly misleading metaphor) to dominate nature, then the prospects for motivating sustainable living on a finite planet are grim. Fortunately, such deterministic arguments rest on untenable assumptions. Let me explain why before wrapping up, since my own position partially rests on the distinction between reductive and nonreductive naturalism.

To his credit, Shepard rejects strong versions of the nature-culture dualism. He takes humans to be natural beings in need of cultural supplementation for survival and flourishing. In his discussion of the ontogenetic life cycle of Pleistocene humans, he claims cultures must supplement biological maturation at all stages of life, since humans are neotenous (in the evolutionary sense of slowed development of the primate pattern and retention of juvenile characteristics).[73] But he also creates the impression that the nature pole of the dichotomy has more determining power than the culture pole.[74] For example, he states that we are "*required by the genome* to proceed along a path of roles, perceptions, performances, understandings, and needs, none of which is specifically detailed by the genome but must be presented by the culture."[75]

This seems to give a large role to culture. Elsewhere, however, he refers to a "hard, irreducible stubborn core of biological urgency, and biological necessity," where "biological *reason*, reserves the right to judge the culture, and resist and revise it."[76] He seems to want to preserve the interdependence of the two poles, but in the end he gives biology precedence as inner determinative power. "We are free to create culture—and have done so in hundreds of ways—but there is a catch.... The catch is that, given a natural world and a human nature, not all cultures work equally well."[77] In the phrase "a natural world and a human nature," the emphasis is apparently placed on some deterministic inner core of human nature rather than on the less deterministic but shared external demands of ecological embeddedness. Pleistocene dreamers invariably prioritize the determinative power of long-term evolutionary (phylogenetic) over ontogenetic events, which implies that a specific kind of cultural adaptation—Pleistocene foraging-hunting culture—best expresses human nature.

With sociobiologists, evolutionary psychologists, and some psychoanalysts, Shepard assumes that what transpired early on in human phylogenetic (and ontogenetic) history is more significant than what has taken place more recently for diagnosing social (or psychological) ills and for prescribing solutions. The current reductionist form of sociobiological explanation of human behavior, evolutionary psychology, proceeds accordingly.[78] The way to understand human behavior is to understand the structure of human brains. The structure of brains is understood by considering the genes or "genetic program" that determines the development of brains. The presence of the genetic program or genes is to be explained only by evolution through natural selection. Because such processes take time, it is assumed that the time period most relevant to explaining modern human genes, brain structure, and behavioral features is the approximately one million years of the late Pleistocene.[79] This seems plausible in broad outline, but when we consider the explanatory relevance of such deterministic accounts, their plausibility evaporates. Everything that we do is something we have evolved the capacity to do, so the claim for evolutionary conditioning is in itself trivial. More importantly, when we seek to explain a behavior we want a proximate reason, and reasons more distal are increasingly less informative. Philosopher of science John Dupré (1952–) argues that the central claim of evolutionary psychology, that genes determine brains and brains determine behavior, is explanatorily empty. This is because "the explanatory relevance of evolution to cognitive mechanisms, and cognitive mechanisms to behavior, does not imply that evolutionary theory has much explanatory relevance to behavior."[80] That is, levels of explanation are not transitive, which runs contrary to the still prevalent reductionist assumption of positivism.[81] In this case, while human phylogeny provides an important interpretive context for human ontogeny, this does not entail that phylogeny has anything directly explanatory to say about human behavior. Primitivists do not attribute equal causal power to the ontogenetic context of development, and yet it is more likely that ontogeny is a more powerful driver of current choices than Stone Age genomic predispositions. Certainly genes play a role in human development, but so do many other causal factors, including, for instance, epigenetic processes (changes in gene expression due to cellular environment), mother's reproductive physiology (stress and diet, among other things), environmental resources either found or constructed by parents (niche construction), and complex parental behavior patterns in offspring rearing, all of which must be reproduced in repeated developmental cycles.[82] Thus, all of the conditions under which brains develop are highly relevant for what results, and so "if we want to know what contemporary human brains are like, reflection on the conditions under which humans (perhaps) lived in the Stone Age is no substitute for the hard empirical work of investigating the nature and variety of contemporary humans." Based on such anti-reductive argumentation, Dupré claims that "brains evolve at the speed of cultural change rather than at the speed of accumulation of genetic change."[83] This view supports the interpretation of human neoteny as nonspecialization to a specific environment, which gives a much larger role to ontogenetic than phylogenetic resources in development and behavior. I explore this point in the next chapter.

At best, primitivist views may be regarded as overcoming the dualistic hyperseparation of humans from nature through an evolutionary reductionism, but at the cost of homogenizing universalization and loss of cultural particularity. They may no longer background nature in philosophical anthropology, but nature is again stereotypically regarded as a blind causal agent determining behavior rather than the ecological context and condition of human growth, reproduction, and development, that is, that on which humans are asymmetrically dependent. Additionally, the implicit naturalistic determinism of these Pleistocene approaches is simply a nonstarter for ethics and politics because it does not address the way that humans can reorient or modify their value priorities in communities, or social ethos, in order to create an environmental culture. These forms of reductive naturalism are thus an inadequate basis for a critical environmentalist philosophical anthropology.

While there are numerous variations on Greek, Panromantic, and naturalistic models, the brief discussions here show that their solutions continue to be shaped by the dualistic Modern conceptual framework. I will sketch a philosophical anthropology that avoids these inadequate responses to dualism in the next chapter.

CHAPTER TWO

The Unfinished Animal

ECOLOGICAL MATERIALISM: EMBEDDEDNESS AND ONTOGENY

We have seen that one of the biggest problems for environmentalist philosophical anthropology is the nature-culture dualism in all of its forms. Dualism not only frames the interpretation of a phenomenon, but frames the interpretation of interpretations themselves. By this I mean that in the academic universe, social constructivists, thought of as one homogeneous group, think of naturalists as one homogeneous group, and treat their explanatory resources, culture and nature, as exclusively disjunctive as well. It is entirely acceptable today for a philosophical critic to submit the *concept* of nature to searching question, but to entirely background questions about what nature really is and why it might matter to see ourselves as material, natural beings. Likewise, those of naturalistic persuasion will try to see human thought and behavior as determined by hidden natural forces, be they brain modules or selfish genes, and to deny and background questions about the historical, cultural, and political meanings and uses of naturalistic argumentation. Reductive naturalists will assimilate culture by explaining it in terms of smallest bits, and constructivists assimilate nature by considering it to be soaked in the dye of cultural meanings. This realism-constructivism debate is plainly just one more crude manifestation of our Cartesian dualistic legacy. As Plumwood herself remarked, this "'two-cultures' division of the field of knowledge into a culture-reductionist humanities versus a nature-reductionist science is a direct contemporary expression of the polarized and dualized choice of nature versus culture characteristic of western culture since classical times."[1]

The tendency of many positions to posit a homogeneous humanity in a clash with a homogeneous nonhuman nature, and to deny its dependence on the nonhuman nature it ostensibly opposes, are moves indisputably linked in the logic of dualism exposed by Plumwood. Rationalist responses perpetuate the dualism because historically rationality has been understood to be a nonnatural feature of humankind, and since in traditional western anthropocentrism universality and rationality have been joined, these accounts tend to regard humanity as independent of nature. A simple reversal of this dualism—the insistence that humans and nature are the same in the case of Panromanticism or naturalism—does nothing to dismantle the dualism itself. A relation of identity or of genealogical descent is simply not the same as a relation of dependence—the latter does not follow from the former. Being products of the same universal evolutionary processes or ontological substrate has no direct bearing on community responses to environmental crisis. The previous chapter has contributed to the overall argument that environmentalist anthropology must involve acknowledging an unamendable structural order of dependence between humans and the natural world, but this idea needs to be further detailed. In this chapter, I will sketch some elements of a philosophical anthropology for critical environmental philosophy that responds more adequately to Plumwood's criteria for a liberation model of anthropocentrism.

Adding to the taxonomy presented in chapter 1, I propose a seventh anthropological model that is broadly naturalistic but nonreductive and nonuniversalizing. Ecological materialism is a view that is suspicious of traditional rationalism, but does not go to the Panromantic extreme of rejecting rationality and civilization entirely. It rejects reductionism, but it does not endorse human exceptionalism either. It embraces and gives a more systematic coherence to the underappreciated nonreductive naturalism that has received considerable treatment by contemporary philosophers such as Marjorie Grene and Mary Midgely, Michael Tomasello and Bruce Wexler in psychology, and the German philosophical anthropologists in the first quarter of the twentieth century. I will draw on some of these sources to highlight some of the features of the approach that respond directly to the problems raised in previous sections, and, looking ahead, establish the basis for the treatment of environmental value theory in the next part of the book. The third section of this chapter already explores some implications of nonreductive naturalism for metaethics. It argues that a central consequence of the breakdown of the nature-culture dualism in anthropology is the breakdown of the needs-wants dualism in ethical theory.

Only two key categories from among the many that could be dealt with here will be the focus of discussion in this chapter. The categories of *embeddedness* and *ontogeny* (developmental life span) will be examined in order to undermine the different forms of dualism that continue to affect most anthropological and environmental theorizing. The concept of embeddedness rejects the universal-particular dualism that underlies universalizing environmentalism, and replaces it with the more nuanced and helpful contrast between variable and shared features of the ecological human condition. The category of ontogeny as used here undermines the nature-culture dualism in environmental anthropologies, and explicitly emphasizes the developmental duration of embodied and embedded human social existence.

The dualism universal and particular generally maps on to nature and culture—universal biology is opposed to particular cultures—and since the category of embeddedness (situatedness) can be used to resist the universal-particular dualism, it can also apply to the nature-culture dualism in places where the two pairs overlap, as in environmental discourses. In place of this dualized pair, I will invoke a distinction between the "variable" and "shared" aspects of the human condition.[2] A familiar nonenvironmentalist example of a discussion of variable and shared aspects of human existence stems from existentialist writings, which rejected universalism and essentialism about human nature in favor of radical individualism. In Sartre's discussion of the human condition, he discusses the pervasiveness of basic moods, absolute freedom of choice, human mortality, being engaged in projects, being with other people, and being in a world only partially of your own choosing.[3] (As environmentalists we have to reject the idealism, humanism, and antinaturalism of a view like Sartre's, of course, which is an extreme form of metaphysical anthropocentrism.) To these shared conditions we should obviously add: eating food, generating waste, burning energy, reproducing, growing and developing in the world with others, and so forth. These conditions are shared and virtually invariable among all human beings, but they do not form a new universal human nature because they cannot be used to explain human behavior in a causally deterministic way, as traditional concepts of human essence or human nature have been used to do. What is shared differs from universality in rejecting determinism, by not attributing *specific* drives, natures, interests, or desires to humankind. The fact of human mortality, for example, does not determine how one responds to that fact; the condition of ecological dependence, the need to eat or generate waste, does not determine what or how one eats or disposes of waste. Likewise, what is variable is not the particular as opposed to universal, since it is the shared conditions in combination with the variation in living them that constitute the situatedness of a person or group. The category of situatedness or embeddedness both includes and contrasts what is invariable and variable in the human condition, like a foreground-background relation in a bas-relief sculpture. A similar conception of this situatedness or embeddedness that is both realist and naturalistic (and slightly less anthropocentric) we owe to Marx.

Marx used the category of *labor* in the sense of everyday human life activity to cut across the classical philosophical divisions between mind and body, human and nature. All human beings "produce their own means of subsistence," though they do this differently in different places, times, and cultures.[4] Here too, the distinction between what is shared and what is variable cannot be transformed into a new universal essence for the human, since the content of these forms remains to be realized concretely by each individual and community in a specific natural-cultural place. If we consider environmental conditions, what is invariable is that food nourishes, water quenches thirst, air sustains, Earth supports and shelters. What is variable is that foods differ, access to quantities and qualities of water varies, air may be polluted or clean, housing and access to land are highly variable even within the same society. Taking up the challenge of environmentalism means recognizing that what might be otherwise passively accepted as universal desires or human interests come to be seen as contingent results of natural-social embeddedness.

Feminist theorists have advanced this issue more than anyone else. The concept of *situatedness* developed by standpoint epistemologists cuts across the categories of universal and particular and is also suitable for addressing the conflict between realist and constructivist views in the sciences and philosophy. Standpoint epistemology is an approach developed by feminist theorists in the late twentieth century in order to counter the traditional associations between universality, objectivity, neutrality, rationality, and masculinity which combined to make it seem as if women (and other dualistically inferiorized groups) were not capable of producing "objective" accounts of the world.[5] In standpoint epistemology, to be "epistemically located" or "situated" is not synonymous with being a "particular," retaining its dualistic associations with subjectivity, emotionality, the body, nature, and so forth. Because situatedness also encompasses what may be shared, semi-invariable structures for whole groups of humans, such as workers or women in certain societies, it bridges the ostensible gap between universal and particular. What is shared is not thereby universal in the classical sense, because again it forms no essence or nature that determines or overrides particularity—what is invariable forms a shared condition which leaves ample room for variable agency, individuality, and initiative. Embodied and embedded subjects are engaged in and through their situatedness, and are more knowledgeably engaged the more they acknowledge the conditions under which they exercise their agency. Natural-cultural agencies have evolutionary and ecologically embodied aspects, as well as sociocultural, historical, and personal ones, some of which may be shared and some variable. For the situated knower or actor, neither nature nor culture can be regarded as a uniform causal power determining behavior any longer. In what follows, I discuss what is shared and variable within the long-term natural-cultural growth, reproduction, and development of human lives in order to better understand the situatedness of environmental agents in their social and political engagements.[6] Below I use the concept of *action* to synthesize these elements of the distinction between the invariable and variable and give a central place to ontogeny, or human development.

I argue that profound human dependence in early life stages captures a kind of relation fundamental to human experience of others and of nonhuman nature. Embeddedness and ontogeny go together—dynamic interchanges with social-natural environments in growth and development are the proximate context of situated agency. Talking about kinds of relations is important because, as we have seen, attempts to describe humans as "part of nature" tend to fall into well-worn patterns of explanation. The "relation" they conceive is one of genetic determination (nature as "origin") or metaphysical identity, and neither is properly "ecological," if we take ecology to be about the interactions and exchanges between organisms and their environments. As paleontologist Niles Eldredge and philosopher Marjorie Grene judiciously noted, both ecological and genealogical processes and relations define living things, and while "neither set of processes is more important than the other in specifying what life actually is," in order to understand human behavior and social systems it is more beneficial to foreground the ecological rather than reproductive dimension.[7] This is another way of saying that talking about the evolutionary timescale in order to account for some set of current ecological behaviors is largely irrelevant. Cultural and developmental psychologist Michael Tomasello makes this distinction in terms of the three timescales within which we have to view human activities. There is the evolutionary or phylogenetic timescale, where biological foundations of current human cognitive capacities developed;

the historical timescale, where cultural products are produced, recorded, retained, and passed on, thereby scaffolding new cultural products; and the ontogenetic timescale, where every individual in the culture develops, and is initiated into the world of social-cognitive intentionality and symbolic, perspectival attention.[8]

Environmentalists usually want to identify the kinds of behaviors that destroy or preserve ecosystems in order to prevent them in the future. If the cultural norms, conventions, and habits of a community are the learning environment of human children from birth, then to pass over the ontogenetic, proximate developmental context of human life and instead reach back to a distant evolutionary past or deep metaphysical identity with nature in order to explain or motivate behavior is misleading. Such accounts "basically ignore all of the social-cultural work that must be done by individuals and groups of individuals, in both historical and ontogenetic time, to create uniquely human cognitive skills and products." Tomasello continues: "Any serious inquiry into human cognition, therefore, must include some account of these historical and ontogenetic processes, which are enabled but not in any way [genetically] determined by human beings' biological adaptation for a special form of social cognition."[9] On this view, culture is not something separate from or tacked on to a biological human after the fact, but is the "species-typical and species-unique 'ontogenetic niche' for human development," and children learn "in, from, and through this environment" from birth.[10] This environment—this real natural-cultural embeddedness—can be conceived as the unreflectively imbibed habitus of a community, where "the particular habitus into which a child is born determines the kinds of social interactions she will have, the kinds of physical objects she will have available, the kinds of learning experiences and opportunities she will encounter, and the kinds of inferences she will draw about the way of life of those around her."[11] In this light, we should have an anthropology for environmentalism that takes ontogeny and human dependence (embeddedness over time) on human-nonhuman surroundings seriously. The shared structures in the developmental context that support variable individual growth and reproduction of human and nonhuman life in communities express these relations of dependence. These structures may be interpreted as necessary conditions under which beings capable of leading a life on their own initiative are reproduced. Thus, embeddedness and lifespan development (ontogeny) are complementary categories. If the proximate cause of current antiecological behavior is cultural patterns of choice, and the proximate context of patterns of choice is ontogeny, then looking for answers to ecological destruction in genes or brain modules is bound to be at best uninformative and at worst a superfluous distraction.

Humans are participants in personal and social worlds whose existence and some of whose dynamics are dependent upon, but not explicable through, biophysical realities. Accordingly, I offer an account of the biological context of human sociality and culture, which does not attempt to reduce culture to nature, but reveals this dependence. More specifically, it shows some of what is shared and what is variable in the situatedness of social ecological agents, forming the backbone of an ecological materialist anthropology.

THE UNFINISHED ANIMAL

Recall that many naturalists contend that human morphology, including the brain, is adapted to life in a specific kind of natural-cultural environment. Based on this premise, primitivists argue that both human and nonhuman nature would be better off if humans adopted the lifeways of our Pleistocene forbearers. From the correct idea of brain plasticity in need of cultural stabilization, they make the (stronger) incorrect inference that humans require a very specific kind of natural-cultural environment (an "uncivilized" one) to flourish, rather than the (weaker) claim that access to some nonhuman nature and cultural scaffolding in general is needed to achieve social and environmental health. Other writers have interpreted this plastic condition of the brain differently, including evolutionary theorists Stephen Jay Gould, Niles Eldredge, and Marjorie Grene. According to these writers, our most conspicuous evolutionary specialization is our nonspecialization, the plasticity of our brains marking, as Arne Naess also noted, humans' "lack of a definite biological place to call home," which he argued forms the basis of the uniquely human capacity to care for nonhumans and the Earth.[12] Notably, German philosophical anthropologist Arnold Gehlen (1904–76) also began his account of human being with this very central insight into human nonspecialization.[13] Drawing primarily on Gehlen's work here, I discuss the "unfinished" nature of humankind and two concepts associated directly with it, action and relief. Many of his ideas have been unwittingly recapitulated by more recent authors, and I will draw out these commonalities. All of them are focused on dismantling the nature-culture dualism in human nature theory, and this is the reason for the similar outcome.

Borrowing a phrase from Friedrich Nietzsche, Gehlen described the human being as "*das noch nicht festgestellte Tier*," the as yet "undetermined" or "unfinished" animal.[14] Taking Nietzsche's pronouncement one way, he argued that the human being is unfinished or indeterminate in the sense of being unspecialized for a given environment, this situation apparently contrasting sharply with that of every other evolved being. "The insight that man [sic] is unspecialized and unadapted to a natural environment and is a 'deficient being' in a morphological sense is of great significance for anthropology."[15] Much later, Grene and Eldredge also remarked on this peculiarity of human anatomy and development: "This is the paradox in human evolution: our most conspicuous specialization is the loss of specialization."[16] Taking Nietzsche's assertion another way, the unfinished human condition may mean that we are a "problem" to ourselves because our indeterminate form provides no clear and determinate paths for our lives to follow. This is meant to contrast with the tight fittedness of other organisms to their environments, which provides many of them with rigid motivational instincts and triggers to release them. Human "instincts" are as unfinished as human morphology, and require education and stabilization. This simple conception of morphological nonspecialization has a surprisingly rich array of implications. First, following a long period of juvenile dependence, a human being is the sort of living being that must not simply survive, but must "lead a life" on its own initiative in a way that differs from that of other animal species. A human being is an acting being. Secondly, it means that humankind has a particular need for sources of stability, such as social institutions, rituals, and habits, which provide structure and "relief" from the gnawing indeterminacy of this condition. It is interesting to note that both properties—purposive acting and lack of clear aims—are themes of existentialist literature and philosophy. Here they are given a naturalistic basis rather than a dualistic, human exceptionalist one.

Gehlen endorsed work in the life sciences of his day that described human morphology as one that shows a regress in specialization, a "retardation" of the primate developmental pattern (which is already itself regressive relative to mammals), and the adoption of an adult form that more closely resembles in structure that of embryonic or juvenile nonhuman primates.[17] The evolutionary process involved is called neoteny. The "retuning" of the primate model due to neoteny "is so radical that it signifies a completely new mode of existence," according to Gehlen.[18] After weighing the evidence, Gould also claimed that human beings are essentially neotenous, and that this "retardation, of itself and apart from any morphological correlates or consequences, has been a factor of paramount importance in human evolution."[19] It is one of the central synergistically acting causes, along with upright posture and expanded brain, leading to the distinctive human condition. Gould continued, "The interacting system of delayed development-upright posture-large brain [is synergistic]: delayed development has produced a large brain by promoting fetal growth rates and has supplied a set of cranial proportions adapted to upright posture. Upright posture freed the hand for tool use and set selection pressures for an expanded brain."[20] This "retarded" development does not necessarily entail nonspecialization, but because in the human case it leads to a much longer window for extrauterine brain development, it logically seems to entail a greater "biological potentiality"—"a brain capable of the full range of human behaviors and rigidity pre-disposed toward none"—as opposed to "biological determinism"—"specific genes for specific behavioral traits."[21] Along with all the paedomorphic characteristics that are evidence for neoteny, Gehlen too presciently argued that the human brain is precisely fitting for a being with such indeterminate morphology because "the brain is the organ that makes any specialization in organ development unnecessary. From the point of view of behavior, the brain is the organ of plasticity, versatility, and adaptability; however, it must always be viewed in conjunction with the unique human constitution—its vulnerability, mobility, and openness to stimuli, as well as its failure to specialize . . . ; the human constitution first makes such a brain possible."[22] Both Gehlen and Gould have described invariable features of the human condition with reference to human morphology and evolution in a nonreductive naturalistic way.

The next step is to take special notice of the long period of extrauterine development that is a developmental consequence of morphological nonspecialization, because it establishes the basis for the unique kind of social learning or cognition characteristic of humankind. Gehlen invoked Swiss zoologist Adolf Portmann's (1897–1982) account to describe the species-typical period of "premature" social learning in human beings. Portmann distinguished between "altricial" mammals with large litters of underdeveloped young (e.g., rabbits) and those "precocial" mammals with small litters of young ready to fend for themselves (e.g., horses). In principle, human beings ought to be classed with their nearest primate relatives as precocial (such as chimpanzees), but they obviously do not fit the pattern. In fact, if it were to follow the typical primate pattern, human gestation would last twenty-one rather than merely nine months. Portmann called the utterly helpless and dependent human infant "secondarily altricial," by which he meant that it combines precocity of sensory apprehension of the world around it with an altricial pattern of dependence. For reasons that probably have to do with the difficulties of passing a large-headed infant through a pelvis designed for upright walking, intrauterine gestation is cut short, and "extrauterine gestation" continues as the infant's brain and morphology continue to develop at embryonic rates in the first year of life. Hence, Portmann called the child's first year its development in the "social uterus."[23] Marjorie Grene also saw that this pattern of development was unique in nature and absolutely necessary for the development of what we would consider a mature human being. Humans reach sexual maturity relatively late in comparison to other species, facilitating a period of growth, development, intensive apprenticeship to the human condition, and "gradual assumption of responsible personhood."[24] Thus, for strictly biological reasons humans grow and develop in a condition of dependency on social and natural conditions not of their own choosing. The nature of this social learning—so important for developing an awareness of human dependency of the environment—will be further discussed below.

Thus, against the broad pattern of highly specialized evolved forms of organic life on Earth, we find that the human form appears to be uniquely unspecialized, and we may draw a number of philosophical conclusions from this. The first bears on the general mode of life of such a being. Gehlen asked first, how, if other beings survive precisely due to their specialization or fittedness for a specific environment, the human being is able to survive at all. "The question logically arises as to the viability of such a being. The theory that *action* forms the center of human existence provides an answer to this question."[25] By "action" Gehlen meant planned, goal-directed behavior. Gehlen's choice of action as a central category is comparable to Marx's choice of "labor." Both set aside the traditional dualistic alternative between mind and body, culture and nature, and instead endorse a "psychophysically neutral" category. For both, action and labor involve muscle and sinew as well as socialized perception, imagination and forethought. The category of *action* is seen to follow directly from the indeterminate, unfinished morphology of the human form. "Man's [sic] disposition toward action is the fundamental structural law behind all human functions and skills, and ... this essential feature clearly results from his physical design. A being with such a physical constitution is viable only as an acting being."[26] He is instructively explicit about the way that this category cuts across traditional dualisms.

> This point of departure unlocks the issue more than any other, since the thinking, knowing, and willing side of humankind is contained in the concept of action as much as its physical side, but such that both are thought

as contained in one another, as reciprocally presupposed, in one and the same act. The empty controversies between biologistic, dualistic, and spiritualistic abstractions are in this manner excluded from the start.[27]

This is a being that does not only live, but must lead a life, orienting itself in ecological and cultural places, prospectively anticipating interactions in a world superabundant with threats and promises, disvalues and values. In this manner, distinctively human action can be distinguished from much nonhuman animal behavior—but we have to guard against traditional dualisms here too. With reference to ethologist Konrad Lorenz, Gehlen distinguished open-ended "orientation responses" from more automatic "instinctual behavior," the former serving as the model for all variable behavior, while the latter is stereotyped and triggered by "releasers." Krogman echoes this distinction decades later, with specific reference to the slowed developmental pattern of humankind: "This long-drawn-out growth period is distinctively human; it makes man [sic] a learning, rather than a purely instinctive, animal. Man [sic] is programmed to *learn* to behave, rather than to react via an imprinted determinative instinctual code."[28] We have to be extremely careful not to underestimate animal learning and rely on another antiquated homogenizing dualism here. Human behavior is also very often automatic and unconscious, even if not a result of what can be called instinct, and nonhuman animal responses can be highly variable and nonstereotyped. We have to dehomogenize within these large classes *human* and *animal* in appropriate contexts of analysis, but the general point about the relation between indeterminate morphology, social learning, and action holds if not taken to a dualistic extreme.

The upshot is that the characterization of humankind as an unfinished animal has its naturalistic basis in evolutionary theory and ecology, but can also be used to guard against going to reductivist or constructivist extremes in our conception of the human condition. The structure of action—value-oriented, indeterminate behavior—follows from the morphologically indeterminate structure of such a being with a plastic brain. This nonreductive naturalistic redescription of the human condition lays the groundwork for avoiding the pitfalls of dualistic accounts. Instead of regarding the human as primarily natural with cultural icing, or primarily cultural with barely any nature left in them, humankind may be productively regarded as "naturally cultural." This brings us to the next important consequence of this developmentalist perspective: morphological indeterminacy also leaves such a being "world-open," subjected to a barrage of information and stimuli demanding to be made sense of. This situation of receptivity to often confusing input puts individuals in a position to seek relief from the problems it presents. Ethically, it puts agents in a position of attempting to resolve felt conflicts of values, and this entails that not all values can be given their due, or that prioritization of some values (goals) above others is obligatory. Before addressing some (meta)ethical implications of this view in the next section, I will further discuss social learning and the concept of relief here.

The emphasis on nonspecialization reveals the importance of developmental context and ontogeny for the utterly dependent human infant in its "openness to the world." Under the broad heading of "communication" or "communicative action," Gehlen drew on the work of Johann Gottfried Herder (1744–1803), Nietzsche, Frederik Buytendijk (1887–1974), and George Herbert Mead (1863–1931) to explore what he called developmental "circular processes" in early human ontogeny. This substantial portion of his analysis constantly drew attention to the intimate relations between our open senses, our corporeal mobility, and the nature of language and language learning. He provided many examples of sensory-motor processes in infants and children that exemplify what he called "communicative acts," or forms of interaction with the world whereby children obtain information about their surroundings by direct sensory-motor interaction with them. He noted how the child's own vocalizations take on a life of their own in processes that exhibit the same circular structure present in other forms of interaction, where the individual is both passive and active at once, both touching the object and being touched by it, both uttering sounds and hearing those sounds outside oneself. This is the basic dynamic structure of all human interaction with things and others in the world, Gehlen argued, and it lies at the root of the important capability of "taking something as something," or adopting a certain perspective on a thing in order to assess its ability to serve as a means to the satisfaction of some ascertained need. Perception does not merely receive impressions or sense data, but learns and anticipates the affordances and capabilities of objects in the process of human communicative action with them. This connection between perception and action was well-mapped later in the twentieth century by ecological psychologist J. J. Gibson, whose work has inspired many in the domain of embodied cognition theory. Marjorie Grene was especially impressed by Gibson's work, and integrated it into her own philosophical anthropology.[29] Adopting a model of perception as value saturated is crucial for environmental philosophers as well because it helps to resolve debilitating problems arising from the dualistic distinctions infecting the discourse of values.[30]

The world presents resistances to human endeavors—it says "no." But it also says "yes," in the sense that the objects in it express "the greatest ontological positivity" through their "affordances"—the possibilities for action given in embodied perception.[31] These affordances are superabundant and polysemic for living, symbol-using beings, and perceiving is as much a matter of limiting the information bombarding the senses by distinguishing what is significant and insignificant in it, as it is simple receptivity. While *affordance* is traditionally defined in terms of physical interaction with worldly objects, for humankind objects also exhibit obvious "possibilities for action" that may be called *values*. In the context of environmental philosophy, Arne Naess criticized the traditional scientific epistemology that segregated primary qualities from all others and argued that we have to begin environmentalist value theory with a revision of our understanding of the place of values in reality. "If we start with designations of concrete contents, for instance 'delicious, red tomato to be eaten at once' or 'repugnant, rotten tomato' the evaluative terms are there from the very beginning of our analysis."[32] On his account, the dualist's problem of explaining how subjective valuations are "projected" onto value-free objects does not arise because he correctly understands the relational, action-oriented, and affectively charged nature of perception. He resisted subjectivism of qualities with reference to relational gestalt psychology (as did other mid-twentieth-century writers), though whether he was entirely successful in this or not remains an open question. We can nevertheless follow his phenomenological impulse and tie it directly to the account of ecological perception in order to avoid the risk of holist subjectivism about qualities. As Naess himself might have said, "What we feel in perception as menacing and attractive is as immediate as redness and greenness as qualities of objects. Only a relatively later reflection teaches us to distinguish between the objective and the subjective in such cases."[33] Better, in everyday perceptual experience we simply take stock of the valuable or harmful qualities that things present to us in the course of our interaction with them. On an embodied view of perception, significant qualities and values of objects are selected out of a cacophonous range of qualities by perception and feeling as affordances over a definite span of lived duration. These qualities tell us what a thing can do and, at first glance, whether it might play a positive or negative role in our ongoing projects. Value qualities serve as "signs of objectively existing relationships, of dangers, threats, opportunities, and the like."[34] These qualities are the "concrete contents" revealed by vital human evaluative capacities, the "mediated immediacy" through which we are in contact with the real world.

It is not only a world buzzing with information to which we are exposed from birth, but a world full of affectively charged attractive and repulsive qualities. This often discordant barrage of qualities has to be handled by the organism struggling to lead a life and flourish in a natural-social world mostly not of its own choosing.

Gehlen claimed that what motivates much human learning is the need to respond to an overstimulating environment and a desire for relief from it. *Relief* is also a surprisingly rich category. The term is used to cover sensory-neural, affective-motivational, and linguistic-conceptual processes. As he described it, the whole process of perception is a process of orientation in a world that provides an overwhelming abundance of stimulation. By actively engaging with the world in a sensory-motor way, the child learns what it and the things it interacts with are capable of, and once it does so it is able to relieve itself of the necessity of doing so with each and every encounter. Children can, eventually, simply by means of a visual sensory overview, ascertain an object's hardness, texture, weight, or what it can do, just by looking at it, and this exemplifies one modality of relief. Other examples of the process of relief in operation include the formation of motor habits; of memory images that come to the aid of current actions when needed; and the reliance on ready-made conceptual frameworks (such as traditional dualisms), stereotypes, basic-level categories, and implicit biases. We cannot live and develop without the scaffolding these many forms of relief often provide. (Some are malleable once we become aware of their implicit operation.) Contemporary cognitive research would add to this by claiming that the developing human brain requires and actively seeks sensory stimulation in order to develop normally, and it must maintain such sensory contact with the world throughout its life. Bruce Wexler reviewed the evidence from sensory deprivation experiments on animals and humans which led to this conclusion. One of the most profound implications of this claim in Wexler's discussion is the principle of internal-external consonance, which says that humans usually try to match their internal neuropsychological structures to aspects of the external environment.[35] Unmatched sets produce anxiety and other sorts of relief-seeking behavior, including strategies such as denial and forgetting of information that does not match expectations or existing internal structures. Thus, to refine Gehlen's claim, it may not be relief from overstimulation per se that is sought, but relief from stimulation that does not match already-existing structures, or avoidance of information for which we do not have ready-made categories or protocols to deal with it.[36]

Psychologist Michael Tomasello explicitly adds to the discussion the important dimension of sociality in these learning processes. While Gehlen discussed ontogeny in a mostly individualistic way, Tomasello emphasizes the social nature of cognition and develops the notion of a shared "intentional affordance." Around the ninth extrauterine month he claims that an important developmental window opens in which some of the other beings surrounding the infant are understood to be "intentional agents like the self."[37] At this point, children engage in situations of "joint attention" with other people, where most cultural learning happens. "Joint attentional scenes are social interactions in which the child and the adult are jointly attending to some third thing, and to one-another's attention to that third thing, for some reasonably extended period of time."[38] Children engage in joint attentional situations with adults, for instance, when adults directly instruct them in the acquisition of a skill or piece of knowledge, or when engaged in play. Their capacity to "put themselves in another's shoes," or imagining them as an intentional agent, is also directly engaged when learning the "intentional affordances of artifacts."[39] In the manipulation of objects, children learn the qualities and properties of objects, and they also learn about their own bodies and senses simultaneously, in a circular process, as Gehlen and researchers like Piaget have noted. Tomasello points to a similar but distinct process through which children watch adults manipulating objects, not just in order to imitate them, but to empathetically insert themselves into the "intentional space" of the user, determine the user's goal, and in this way see what the object can do, or what it is for. So, in addition to the sensory properties of the object, it will be seen to have "another set of what we might call intentional affordances based on her understanding of the intentional relations that other persons have with that object or artifact—that is, the intentional relations that other people have to the world through the artifact."[40] Tomasello emphasizes the importance of such interactive learning and social cognition during the long apprenticeship needed to hone social-cognitive skills in the ontogenetic context for a being with indeterminate morphology. In social life, such affordances can tell us not only what can be done with others but what should be done with them, creating what we could call "normative affordances."

These are just a few of the implications of adopting an ecological materialism, or nonreductive naturalistic approach. This kind of naturalism avoids reduction by taking dependence and ontogeny seriously. By accepting an evolutionary point of departure, it affirms that humankind is part of nature, a product of the same processes of organic evolution in operation throughout nature. At the same time, it affirms human distinctiveness with a nonhierarchical conception of human difference, relying on description of evolutionary processes to highlight humankind's distinct way of life. It does not dive down to the level of genes, reach far back into the evolutionary past, or make recourse to some metaphysical identity in order to explain human action. It keeps ontogeny squarely in view. This approach to human uniqueness has clear implications for the way we think about the nature-culture dualism. All of the structures necessary for the performance of cultural feats, such as larger brains, opposable thumbs, laryngeal development, and pelvic shaping, begin to form embryonically. That is, the *natural* anatomical and developmental conditions of human life form the supporting structure for the *cultural* human condition, but cultural affordances also provide the immediate environment for these natural potentialities to develop. Since without culture we would lack a "human nature" altogether, it seems reasonable to conclude that *culture is natural* for human beings. Culture—including symbolic perception and language, institutionalized social structure, shared norms and practices—is not something added on to a creature that could otherwise exist without it. As Grene stated long ago,

> The whole structure of the embryo, the whole rhythm of growth, appears to be directed to the emergence of a culture-dwelling animal—an animal not bound within a predetermined ecological niche like the tern or the stag or the dragonfly or even the chimpanzee but, in its very tissues and organs and aptitudes, born to be *open to its world*, or, better, open to a world within a world, to be able to accept responsibility, to make the traditions of a historical past its own, and to remake them into an unforeseeable future.[41]

Gehlen concurs with Grene here when he insisted that "human being is a cultural being by nature."[42] Like Gehlen, Grene, Plumwood, and others, Tomasello argues that "there is thus no question of opposing nature versus nurture; nurture is just one of the many forms that nature may take."[43] It appears that culture depends upon, and is intimately interwoven with, the existence of a being with indeterminate morphology and indeterminate impulses that must be cultivated in light of extensive experiences in natural and social worlds, impulses that are made concrete in the institutions or second nature, within which human beings grow to maturity. It is important to note that where the culture-nature dualism is dismantled, the result is not some kind of unity or identity of the terms. Structural relations of dependence exist between them, and these relations between vital and cultural-symbolic "strata" of the real world will be discussed in the final chapter.

To sum up, embeddedness or situatedness is the invariable natural-cultural condition of humankind, featuring unspecialized morphology, a long period of utter dependency on caregivers, developmental plasticity, circular learning processes and social cognition. These entail openness to the world, or embodied perception of an abundance of value qualities, and initiative taken (action) to orient, relieve, and regulate the overwhelming influx. The concept of the unfinished animal not only affirms both continuity and distinctiveness, avoids reductionism, and eliminates a number of pesky dualisms, it also provides a naturalistic metaethical foundation for environmental value theory.

ACTION, ARTICULATION, AND VALUE

As might be expected, environmental value theory has often implicitly rested on a dualistic conception of motivation as well. Rational aims are opposed to emotional desires, cultural interests to biological needs. Recalling Paul Taylor's metaethical conceptions above, he claimed that environmental ethics is "an attempt to establish the rational grounds for a system of moral principles by which human treatment of natural ecosystems and their wild communities of life ought to be guided," rather than, for instance, a consideration of the role of sympathetic or caring responses to nonhumans in moral life.[44] Emotions such as care and love, and virtues such as compassion and sympathy, generally take a back seat to rules and abstract principles that are to apply the same way for everyone, everywhere. Approaches like this presuppose a model of the primarily rational human being that ecological materialist anthropology rejects. Policy proposals also often rest on these inadequate models. Wildlife managers guided by the criterion of maximum sustainable yield are most likely not allowing their decisions to be informed by their affective responses to ecosystems and nonhuman animals, to put it charitably. In order to approach environmental values questions more productively, we have to explore the implications of this ecological materialist anthropology for value theory as well. Since overcoming the metaethical forms of the culture-nature dualism will be unfamiliar, I once again direct attention to Gehlen's treatment of the issue here. Once the connections are made, however, it will be seen to be an obvious implication of the theory developed, and more of its details will be further worked out in subsequent chapters.

The kind of being that must ultimately act on its own initiative and lead a life is a being whose motivations are as coarsely defined as its unspecialized morphology. To the indeterminate (unfinished) nature of the human body corresponds a rough and indeterminate "excess of impulses" with no ready-made or deeply channeled paths of discharge. According to Gehlen, this "*constitutional* excess of impulses" is "the internal aspect of an unspecialized, organically helpless being who suffers chronic pressure from internal and external stresses. The excessive impulses are thus a reflection of man's [sic] chronic state of need."[45] Since the indeterminate human form arises in a world of change, challenge, and unpredictability, the human problem is both to cultivate and direct its mostly undifferentiated energies in directions that can respond to these changes, and to keep them constant despite potential social and environmental changes in order "to resume an activity the following day."[46] The surplus of impulses can also be thought of conversely as a loss or "reduction of instincts," or as a "dedifferentiation" of drive energies.[47] As a consequence of this excess and indeterminacy, human needs or interests have to be defined in the course of action in the world, that is, naturally-culturally.[48] These impulses are initially quite amorphous and first gain form through the accumulation of experiences and are further developed through experience. Impulses are "plastic and variable," and adjust to changes in experience and situation, and even grow from repeated actions themselves. Therefore, because the character of "needs and impulses" depends upon action and experience, "there is no sharp boundary between elemental needs and qualified interests." The line between need and interest (want, desire) is far less distinct than most theories of human motivation lead us to suppose. In other words, as another consequence of starting with indeterminate morphology, this account entirely blurs the dualism between supposedly essential, biological needs and superfluous, cultural wants assumed by many ethical and political theories.

Gehlen used "the terms impulses, needs, and interests synonymously and always in plural form."[49] An ecological materialist anthropology should accept this as a fundamental corollary for metaethics. Given what has also been said about the cultural shaping of the plastic brain in ontogeny, it becomes apparent that there can be no firm line between biological needs and cultural interests.[50] Every mode of action or production for humankind is naturally cultural, and "culture is therefore more than simply reasonable for man; it is in fact vitally necessary to his survival."[51] Human impulses are thus "developed and formed" in a changing cultural and natural context, and can "arise from actions themselves which then become needs." This is frequently found when the performance of an action that was once the means for the satisfaction of one need or interest can itself become the object and aim of action, and a new need has arisen, a need seemingly as compelling as the first. For instance, every day in consumer capitalist societies advertisers bank on this mechanism and the inherent variability of needs.[52] This plasticity and on-the-fly redefinition of needs is a necessary characteristic of human life in a changing world—the assumption being that an entity with fixed instincts could only survive in a fixed environment. "For an acting being, a being who is exposed to the unrestricted randomness of reality, it is crucial that even the most refined abilities be able to become needs and to be executed with great interest."[53] For example, we often speak of "needing to go to work." Objectively speaking, no one *needs* to go to the factory to perform deadening labor, or to the university to teach classes, or to the stock exchange to supply the financial system with impetus for running the capital mills one more day; but performing these sophisticated actions comes to feel as necessary as any biological need one has ever experienced—or even more so. The value of work is natural-cultural in the sense that through ontogenetic processes it is recorded in the synapses of developing brains, the social-natural ecology of the environment, and in the narratives and institutions of the environing culture. According to Wexler, "our brains (and minds) develop concrete perceptual structures, capabilities, and sensitivities based on prominent features of the environment in which we are reared, and then are more able and more likely to see those features in the sensory mix of new environments we encounter."[54] These sensitivities and value responses are quite literally *inscribed* in the structures of the brain, which depends on social scaffolding for normal development.[55] An environmental implication is that humans appear to be quite satisfied—even if often not very healthy—reproducing the environments they grew up in. If we continue to raise children in crowded cities where environmental values have the lowest priority, they will grow up to be adults who prefer crowded cities and ignore environmental problems.[56] This is not a situation that more scientific information and well-intentioned arguments over the nature of intrinsic value will resolve on their own—it is a matter of producing and reproducing an environmental culture for a mixed human and nonhuman collective. Producing, reproducing, and developing human life in an environmental culture requires that we dispense with the idea of fixed needs and instead adopt the principle that our needs "be developed in close relationship with active experience, and that they become, without sharp demarcation, interests in specific situations and activities.... On the other hand, enduring interests must also be cultivated, oriented, and retained and must remain conscious as inner invariants which control and outlast any changes in activities and circumstances of the present."[57] One of the most important problems we face in responding to mostly self-induced environmental challenges is "organizing this architectonic and well-oriented system of impulses." This has far-reaching repercussions for our theories of environmental value.

In other words, as long as we mistakenly divide human life into a need-burdened, objective biological half and a want-driven, subjective cultural half we will be unable to overcome environmentally inflected versions of the relativism-universalism debates. Instead of worrying over the poorly posed question of where to draw the line, of definitively answering the questions "what needs are transculturally universal?" and "what interests are culturally relative and so not fundamental?" we are faced with a different problem and a single question: "what should our priorities be?" As I argue below and in the following chapters, these priorities are articulated in the language of values. Eliminating the line between needs and wants removes an obstacle to having a single language of values, and a single language of values is needed because it allows us to talk about priorities directly—that is, it allows us to resolve conflicts between competing values in reproducible ways that inform habits and patterns of prioritizing.

As language-using animals, we must recognize the significant role that linguistic expression plays in motivation. Organizing this system of impulses is a largely linguistic affair, and is effected by naming our emotional responses, creating value terms, and putting them into categories of *need* or *want* (among others) by cultural convention. If we understand human language naturalistically as an organ of perception that reveals qualities and affordances in the world just as do eyes, ears, and hands, then the close connection between language and (value) perception becomes clearer. Tomasello devoted attention to linguistic communication with an eye to revealing not only the dialogical possibilities implicit in language learning contexts, but also the development of symbolic representation and what he calls "representational redescription." A revealing discussion of early linguistic development shows that as children acquire language, a kind of obligatory perspectivism is built into language itself, since often the same object can be designated (for example) a flower, a rose, a plant, a gift, or a sign of love, which requires the hearer to ascertain the speaker's meaning from among a host of possibilities.[58] Tomasello uses the term *perspective* to mean "the possibility of placing the same entity into different conceptual categories for different communicative or other purposes."[59] Examples of such situations abound: "In different communicative situations one and the same object may be construed as a dog, an animal, a pet, or a pest; one and the same event may be construed as running, moving, fleeing, or surviving; one and the same place may be construed as the coast, the shore, the beach, or the sand—all depending on the communicative goals of the speaker."[60] As language learners then, children also "acquire the ability to adopt multiple perspectives simultaneously on one and the same perceptual situation."[61] Gehlen also remarked that language enriches our perception of the world by introducing new perspectives.[62] Tomasello takes this notion further by arguing that language itself enables "multiple simultaneous representations of each and every, indeed all possible, perceptual situations."[63] The significance of this ability cannot be overestimated, and in conjunction with the notion of intentional affordances (interactive properties), amounts to an imaginative capacity to evaluate aspects of our environments in nuanced ways. These perspectives communicate substantive qualities of things nonmoral and moral. Something is strong, or hard, or brittle, or alive, or nutritious, or heat retaining; or persons are just, or generous, or courageous, or kind, and all of these descriptions are at the same time evaluations within the context of action and experience. This kind of cognition is also clearly social, since it is often the case that "explicitly symbolized perspectives of interactants clash and so must be negotiated and resolved."[64] This clashing of perspectives is perhaps one root of our experience of value conflict.

To his great credit, by beginning with the notion of an "excess of impulses" in the human being, Gehlen found it self-evident that human beings are a conflicted species.[65] By centering discussion firmly in the concept of action, we may avoid a dualistic casting of these conflicts in terms of reason versus inclination, in any of their traditional forms. In leading a life it becomes one of the central tasks of the human being to orient and organize these impulses in the midst of communities with already-existing conventional and institutional pathways of expression for these impulses.[66]

Thus, human impulses require some degree of cultivation and description in order to be explicitly recognized, and humans constantly engage in ordering and organizing a hierarchy of values that becomes a regular and automatic pattern of impulses and action. (I call this the "problem of prioritization" in part 2 of the book.) We can call these moments of evaluative expression "articulation." Articulations which contrast one kind of value with another, with all the cognitive and affective charges and practical consequences they entail, are essential to clarifying our own motivations both to ourselves, to others, and for any collective planning processes. Charles Taylor's expressivist theory of language, also inspired by Herder, holds that articulation of values is an essential aspect of being a moral agent.[67] The role that language plays in awareness of our own motivations is indispensable on Taylor's theory of motivation. "If language serves to express/realize a new kind of awareness; then it may not only make possible a new awareness of things, an ability to describe them; but also new ways of feeling, of responding to things. If in expressing our thoughts about things, we can come to have new thoughts; then in expressing our feelings, we can come to have transformed feelings."[68] Likewise, since there is a connection between our feelings and values (terms of qualitative contrast), it follows that by expressing our values we may come to have transformed values and motivations as well. This kind of individual or social reflection not only pertains to isolated feelings and responses, but to a larger self-understanding or whole mode of life. "To characterize one desire or inclination as worthier, or nobler, or more integrated, etc., than others is to speak of it in terms of the kind of quality of life which it expresses and sustains, … the kind of life and kind of subject that these desires properly belong to."[69] Because we are self-interpreting language-using animals, our formulations (as well as others') shape our experiences as much as our experiences shape our articulations, by allowing us to discern more, more subtly, and to make manifest things only language is capable of manifesting.[70] The description of a valuational situation that highlights significant value-charged aspects (such as the goods values present in it) allows the agent to reflect on their intentions at the second order. For example, a child can complain to the teacher that her distribution of textbooks was not fair because the one the child received was missing a cover while all the other children received one with cover intact. The teacher can reflect on whether her act was fair or not by highlighting that covers are not important, and what is important is that it is the same textbook (in terms of content) for everyone. There can be disagreements about articulations, and there is no single best account. We have to recognize the fallibility of our articulations and enlist the judgment of others. When we discuss what is significant or important to us we have culturally delimited languages of qualitative contrasts, namely, value terms, which allow us to orient ourselves in an otherwise over-full, value-suffused perceptual and experiential reality. Linguistic articulation is essential for relieving this burden of stimulation and for bringing order to the excess of impulses by naming their contents with specific, qualitative value terms; for manifesting the multiplicity of values in the same situation or object; and for transforming vision and sensitivity in the midst of social engagement. Planning, orientation, and strategy all require implicit reference to what may be called values or clusters of values. Different cultures codify and institutionalize values or value clusters differently, but all cultures present a range of implicit values (inscribed in brains and the affordances of things and relations) that then become amenable to individual or community prioritization.

In part 2 the problem of organizing and prioritizing values is given a central place. A more fully developed discourse of values is indispensable to projects of social change because human beings are the kind of animal that must orient its strivings in a world suffused with articulable values that have to be prioritized. I will argue that a discourse of values is not only best suited to revealing the order and organization of this domain, but that it is essential for environmentalists who must articulate a specific configuration of environmentalist values in contrast to dominant anthropocentric ones. While many environmental philosophers are right to foreground terms like *instrumental* and *intrinsic value* in contrast to other available ethical terms like *rights* or *duties*, these designations are not enough for the task of prioritizing, as I explain in the next chapter. The approach rests on the insight, however, that a discourse of values is somehow more basic than any other. This is because virtues, duties, rights, principles, rules, and commandments can be explained with reference to something that ought to be, a compelling attractor integrated into the nexus of purposive action. All ethicists implicitly or explicitly claim that virtues, duties, rights, and so forth ought to be. However, the meaning of *ought* can only be effectively resolved in the notion of worth or value. What it means for something to be valuable is that it ought to be, and for something to be a disvalue means it ought not to be.[71] Not all values have an equally strong claim on us, or normative pull, and we have to differentiate between types of values. But there is no more basic explanatory term than *value* for understanding the normativity of norms. The consequences of this view will be elaborated in the following chapters.

This brief sketch of some elements for an ecological materialist anthropology—a seventh anthropological model for environmentalism—makes it clear that it is not only necessary to engage with philosophical anthropology in the context of environmentalism, it is necessary to be on the lookout for dualistic formulations even in those cases where the crudest nature-culture dualism has apparently been subverted. The model laid out here fulfills Plumwood's criteria for a nonhierarchical (nonseparatist), dehomogenizing view that acknowledges natural and social forms of dependence. Defining nonhumans in (nonassimilating) positive terms and noninstrumental valuing will be discussed in part 2. What has been shown here is that it is not only nature-culture, but mind-body, reason-emotion, and need-interest dualisms that must be critically assessed. The concept of the unfinished animal and its emphasis on ontogeny and embeddedness provides a nonreductive naturalistic model that undermines the nature-culture dualism, and the principle of action entails that indeterminate motives are in need of experience and cultivation, defusing the long-standing dualism between needs and interests, or biology and culture, in moral psychology. The principle of dependence is respected by acknowledging that agents can only orient themselves at all when a preexisting natural and cultural world to which they must adapt is presupposed. Orienting and structuring amorphous impulses becomes a major task for this kind of being. Seeking relief through stabilization—linguistic categorization, habit formation, institutions—is a second profound consequence. Since linguistic expression functions as an organ of perception, articulating what its needs or interests (values) are plays a central role in orienting action of such a being. Centering on the ontogenetic temporal scale, Gehlen, Wexler, and Tomasello reveal processes through which children come to structure their perception and motivations, specifically through imitation, language learning, and joint attentional processes. Mentally and empathetically occupying the place of the other in order to find out what things and others can do reveals the affordances, or interactive value qualities, that things and others have for potential (inter)action. These ontogenetic processes are crucial for the formation of stable, culturally informed, and value-motivated action patterns. If the environmental crisis can be characterized as a problem of cultural misprioritization conditioned by stereotyped dualistic conceptions of nature and powered by social institutions, narratives, and habitual practices that reify harmful patterns of prioritization, we need to not only avoid dualisms in our conception of the moral agent, but must also be aware of the way that dominant dualistic cultural narratives shape moral reasoning itself. The anthropological and metaethical groundwork laid in this chapter will be put to use in part 2 on the second major topic in critical environmental philosophy—the intrinsic value of nature.

PART II

The Intrinsic Value of Nature

CHAPTER THREE

The Problem of Intrinsic Value and the Primacy of Priorities

INTRINSIC VALUE AND METAETHICS

Perhaps even more than the critique of anthropocentrism, environmentalists' claims about the intrinsic value of nonhuman nature communicate the powerful impression that humans inhabit a world not made for us. Against the backdrop of the Modern conceptual framework, however, such a claim seems just as powerfully counterintuitive to many people. It is no coincidence that there is a great deal of conceptual convergence in Modern epistemology, value theory, and political economic theory.[1] They share the assumption that human thought, desire, or activity create value and order in the world as we experience it, thereby humanizing it—if not creating it from whole cloth. The world appears to us the way it does because of the categories humans use to think it (Kant); it has the value it does because of human interests in it or desire for parts of it for their own use (Hume); things and land have the economic value they have because of the labor time humans invest working on or occupying them (Locke, Marx). These correspondences are all part of the generalized anthropocentric philosophy of Enlightenment humanism, built around the dualistic logic discussed in part 1 of the book. Its bottom line is that the world is made *for us* because, at least in part, it is made *by us* in a priori fashion, and cut to the measure of our categories, values, and desires. (This is the more subtle and metaphysically "modest" version of the ancient, explicitly teleological claim that the world was made for human beings by a divine creator.) Once this anthropocentric order is taken for granted, it becomes a challenge to conceive of "things as they are in themselves," without human contribution, let alone what intrinsic value such things might have independent of human cognition, desire, or activity. Modernist anthropocentrism falsifies the facts of embeddedness and ontogeny covered in the previous chapter by denying the dependence of the human on the more-than-human world. Intrinsic value theory in environmental ethics at its best should be seen as an attempt to refute the subjectivist or anthropocentric trend in value theory.

However, its promising start was crippled from the outset by the Modernist way in which the problem of intrinsic value was framed. The literature on intrinsic value is vast, and all I can do here is outline the Modernist frame of the debate to serve as background for the value theory developed in this part of the book. It was widely believed (even if not explicitly stated) that development of a normative theory of environmental ethics could not begin in earnest until the epistemology and ontology of intrinsic value was settled. The contrast between instrumental and intrinsic value was widely used, although the term intrinsic value was often just as ambiguous as the term anthropocentrism. By the early 1990s, as John O'Neill noted, it was time to take stock of "the varieties of intrinsic value" on offer, and to consider which was indeed central to environmental ethics.[2] O'Neill helpfully distinguishes four distinct senses of intrinsic value employed in the literature, three of which I refer to here.[3] Intrinsic value was (1) used as a synonym for "noninstrumental" value. If "instrumental value" is what something has by virtue of being a means to some end, noninstrumental value consists in being an ultimate end without further reference. This draws on the familiar Aristotelian point that chains of means and ends have to lead to an ultimate end somewhere, so some objects, states, or activities have to be understood as ends in themselves, or as intrinsically valuable. (2) The second sense of intrinsic value is Kantian. Because a central feature of environmental ethics has been the attempt to extend "moral considerability" to nonhumans, appeals to intrinsic value were also made in the sense of "end in itself" that Kant employed. Persons as ends in themselves are moral agents and members of the "realm of ends." If a tree has intrinsic value in this sense, then it is a member of the moral community, is "morally considerable," a "rights-holder," or "has standing." (3) Finally, intrinsic value may also be used as a synonym for objective value, meaning the value that a thing possesses independently of the minds and evaluations of valuers. Meaning (3) expresses the strongest metaethical moral realist claim about the nature of intrinsic value. O'Neill thought that any environmental ethics seems obliged to attribute value or intrinsic value to nonhumans in at least one of the first two senses. The central claim of nonanthropocentric ethics is that nonhumans are not valuable merely because they serve as means to human ends, but are valued for their own sake, are morally considerable, or have "a good of their own." It remained an open question whether this value must also be understood in terms of the stronger sense in order to establish an environmental ethic. Since these senses were often conflated in the literature, the general impression was (and usually still is) that intrinsic value in one of the first two senses entails objective value in sense (3). The debate was for the most part not over which of these senses is correct, best, or most productive for environmental ethics, but whether value was subjective or objective. In other words, it was entirely bogged down in dualism.

My aim here is not to review the debate as a whole or cast judgment on it, but to simply note that its terms were set in advance by the Modern constitution. Utilizing its conceptual resources—like the distinction between subjective and objective—without significant modification, means that no matter how well-intentioned they are the responses to problems perceived through its lens will be inevitably shaped by it.[4] I refer primarily to sense (3) here, leaving a discussion of sense (2) to the next section. To simplify a complex debate, we could say that J. Baird Callicott endorsed a subjectivism of value, while Holmes Rolston emphasized the subject-independence of at least some forms of value. They disagreed on the question of the relation of nonhuman value to a human valuer. According to Callicott's Humean model, such a thing as intrinsic human-independent value cannot exist at all. His distinction between intrinsic and inherent value makes this apparent. "Let something be said to possess *intrinsic* value, on the one hand, if its value is objective and independent of all valuing consciousness. On the other, let something be said to possess *inherent* value, if (while its value is not independent of all valuing consciousness) it is valued for itself and not only and merely because it serves as a means to satisfy the desires, further the interests, or occasion the preferred experiences of the valuers."[5] He thus rejects intrinsic value in the objective sense (3). Value is never independent of a valuer, but this does not mean, he is careful to note, that subject-dependent value is always self-interested value. Value depends on valuers, but the valuing intention itself may have its ultimate target in the object or content, with no further reference back to the self or its desires. "Valuing subjects for the most part do value themselves, but they may, and very often do, value other things equally with or even more highly than themselves."[6] However, sourcing value in the subject, even if locating it occasionally outside the subject, cannot lead to a satisfactory solution to the larger relativist dilemma posed by the anthropocentric problem of exploitation of nature discussed in the introduction. Some individuals will exploit nature because they do not inherently value it; some will not because they do. Inherent value in this sense cannot provide an objective anchor for value in nonhuman nature since its source remains entirely internal to the valuing consciousness. Even if, on a more charitable reading, we agree that whole social groups can inherently value nature, the Modernist categories "subjective" (particular) and "objective" (universal) fail to capture the ontological and epistemological status of such social entities.

Rolston, on the other hand, often used *intrinsic* in sense (3) of independent rather than merely noninstrumental value. He claimed that "there are values that only come with consciousness, but it does not follow that consciousness, when it brings new values, confers all value and discovers none."[7] In contrast to Callicott, Rolston tenaciously argued that for the sake of environmentalism it is important to resist subjectivism in all its forms, and that there are nonhuman valuers—and hence nonhuman value. "What is really incredible is that we humans, arriving late on the evolutionary scene, ourselves products of it, bring all the value into the world, when and as we turn our attention to our sources. That claim has too much subjective bias. It values a late product of the system, psychological life, and subordinates everything else to this."[8] Instead, he asserts, "nature is not barren of value; it is rather the bearer of value."[9] He claimed that at least some value in the world is independent of anthropogenic valuing. He identified "intrinsic natural value" with the real existence of "valuers" in the form of plants, animals, and species as historical entities, and of value-producers in the form of whole communities, ecosystems, and even landforms. Perhaps his clearest case for this is that of nonhuman animals. "There is no better evidence of non-human values and valuers than spontaneous wildlife, born free and on its own. Animals hunt and howl, find shelter, seek out their habitats and mates, care for their young, flee from threats, grow hungry, thirsty, hot, tired, excited, sleepy. They suffer injury and lick their wounds. Here we are quite convinced that value is non-anthropogenic, to say nothing of anthropocentric."[10] Therefore, on the evidence that beings other than humans value themselves and others, we have to admit that value exists independent of human value attribution. Similarly, arguing on Aristotelian grounds, O'Neill held that the "strong objectivity" of value may be based in an understanding of the flourishing of a being, for whom things can do good or ill. In such cases, what is valuable for the welfare of a being can often be characterized completely independently of the evaluations of that being, or of an observer. "In order to characterize the conditions which are constitutive of the flourishing of a living thing one need make no reference to the experiences of human observers."[11] Rolston's conception of the intrinsic value of animal life may also be conceived in this way. Such life thus has value independent of human valuation based on the idea that it values itself, and this may be all environmentalists need to build their "new" ethics. Even Tom Regan's concept of being "subject of a life" and Paul Taylor's of a being's "having a good of its own" can be interpreted in this light, implicitly including the sense that these beings are "self-valuers."

There is far more detail and nuance to these debates than I can discuss here. In the literature, authors such as Keekok Lee, Anthony Weston, and Bryan Norton, among others, criticized or attempted to reform the concepts and clarify the framework in question. Lee attempted to reconcile the two positions.[12] Weston criticized the still nascent quest for intrinsic value in nature and also laid the groundwork for an alternative.[13] Norton drew the helpful distinction between epistemic and ontological aspects of the problem. The epistemic question whether environmentalists' claims about values are "epistemically justifiable" is important and has to be answered, but whether environmentalists' values are located "'out there' in the world itself, independent of human consciousness" is simply a bad question, he argues, and is the product of an outmoded representationalist and foundationalist conception of epistemology.[14]

Ultimately, what this debate reveals most palpably is its Modernist starting point. Recall that the quest for intrinsic value (particularly in sense 3) was an attempt to satisfy a perceived need in the context of late twentieth-century culture, namely, the need for a transcultural anchor of value. It is framed by the same Modernist dualism that motivates the issue of cultural relativism, since rights for nature would seem to be guaranteed if something acultural, transhistorical, and independent of human desire could be found. Norton put it this way: "Commitment to self-evident rights in nature would provide environmentalists with an objectively discoverable, logically supportable, and culturally independent moral talisman."[15] Against these universality and independence assumptions, pragmatists like Norton rightly acknowledged that "inherent value is attributed within cultures, experienced within a historical and intellectual context, not *ex nihilo*."[16] Here all that is needed is a conception of intrinsic value in the sense of noninstrumental value (1). For them, the ultimate units of value theory are contrasting, qualitative values in specific contexts, such as ecosystem health, biodiversity, integrity, beauty, wildness, and resilience, for example, rather than sense (3) intrinsic value. Against the Modernist coding, then, they suggest that culture should not be taken as a metaphysical extension of the subjective, but as an objective social and institutional reality in the context of growth, reproduction, and survival (which requires modification of the Modernist dualist ontology):

> There are cultural values that are the product of social evolution. These values are not entirely subjective. At any given moment in the history of a particular society they can be objectively identified and described. Moreover, in most cases they are the foundation for the values of individual people. It is no accident that nearly all people in a particular society share the same values. They pick them up as children without formal teaching. They are the context and starting point out of which individual differences develop. Simply to call these social values subjective misrepresents their very substantial objective character.[17]

The consequence of this conception of noninstrumental value is the immediate obsolescence of the term *intrinsic value*, at least in sense (3). The two categories of instrumental and intrinsic value then turn out to be empty placeholders for concrete value contents that remain to be specified. The reification of these abstractions was misleading. This pragmatist critique of the concept of intrinsic value and its substitution by a more anthropologically adequate view is consistent with the ecological materialist anthropology outlined in the previous chapter. I'll show how it can be taken further in the next chapter. I discuss other aspects of the environmental pragmatist challenge in section 3 below.

There is another feature of the statement of the problem of intrinsic value, aside from its dualism, attention to which reveals a much more serious challenge to its viability. Arguments for the existence of intrinsic value in nature were carried out against the backdrop of a Modern concept of nature imagined to be empty of all meaning and value until human beings or analogous valuers appear on the scene. Given the pervasive impact of Cartesian dualism, the apparent aim was to demonstrate (in face of skepticism, subjectivism, and relativism) that nature has an intrinsic value in the strong sense, maximally and exclusively independent of the awareness of an observer. However, even if the arguments for the independent existence of intrinsic value are successful, they still do not show that such beings' claims on us deserve less, equal, or even more consideration than we would give to other humans' claims. In other words, although it could conceivably be established that living things and ecosystems are in the moral ballpark or possess intrinsic value, we are still left to decide whose claim is more important, or which interest overrides which other in cases of conflict. This is not a trivial moral problem. It is this question of prioritization of claims or values which plagues individuals and communities in our everyday interactions with nonhuman nature, even when conflicts are resolved through habit or custom more than through participatory engagement. The most urgent values question, it might be argued, is the axiological one concerning which value claims take priority in cases of conflict. In anthropocentric instrumentalization, for instance, what is at stake is the differential preference given to human values, not whether nonhumans have or do not have value at all. To say that things are useful for some end is also a general way of claiming that things have value. The instrumental-intrinsic distinction abstracted from this experiential situation of value conflict has distracted environmental ethics from its primary task for too long. The thrust of the pragmatist and feminist critiques of these debates is that we have to return to the context of moral experience from which the need for a concept of intrinsic value arises. This is the experience of moral conflict. In the last chapter I claimed that one of the most important tasks for an unfinished animal is ordering and organizing its indeterminate impulses. This means that we have to take the concept of action as a point of departure when reckoning with an unfinished being and its polymorphous motivational experience. The category of action already implies a world of abundant values and disvalues that plurally and prospectively orient human behavior. A crucial task for environmental ethics and politics is creating a framework for understanding value priorities and how to reorient them within the already existing field of value experience. I will lay out my own theoretical way into this field of experience, which begins with the distinction between "property" and "priority" questions. The problem of prioritization arises from the pervasive experience of value conflict for an acting being.

PROPERTY AND PRIORITY QUESTIONS

Approaches to ethics can be distinguished in terms of their overriding concern with the "properties" of moral agents and patients or with action-orienting value "priorities." Possession of certain essential properties endows beings with intrinsic value in sense (2), granting them moral considerability. Ethicists have argued that if a being is sentient, has "a good of its own," is "subject of a life," is a person, has dignity, or a "face," it is entitled to inclusion in the moral club. Approaches to ethics which take the taxonomic task of identifying a class of such "morally considerable" entities and their properties to be the most important one for ethics I call *propertarian*. The propertarian approach is driven by the common metaphysical-ethical assumption that it is the constitution of a being—its special properties—that should primarily determine our moral responses to it. Asserting the intrinsic value of an entity is another way to do this. On this model, it seems logically necessary to know *what* something is before we can respond appropriately, so creating the right ethical taxonomy is the decisive issue. Folk taxonomies, stories, and anthropocentric narratives already shape our moral responses. The categories at play in ethical theory, such as *person* or *human being*, never stand alone, but have meaning only in contrast to other basic categories, such as *animal* or *thing*. Propertarian tendencies are a specific kind of response to the principle of relief explored in the last chapter, since we want to know what something is (what it can do) in order to decide whether to expend effort on it or not. Given our limited resources, not everything can get our full attention, so we have to make choices. Under triage conditions we may simply seek relief, and carve complex situations into two-valued, exclusive alternatives. There are obviously many cases where this kind of oversimplifying categorization—which aims to divide the world or entities into two classes—is *too* relieving, and becomes a convenient ideological falsification of phenomena experienced. I therefore suggest that ethical theory should not begin with theoretical propertarian categories like rights, intrinsic value, sentience, face, and so forth, because they are simplifying abstractions from a much richer field of experiential conflicts of values in embodied-embedded life contexts. We need an intermediate level of complexity in our classificatory scheme to facilitate better contextualized and more nuanced study and response. While there are obvious differences between approaches that emphasize rights-talk or sentience, for example, and these are clearly very different ethical categories, the point here is that they both obscure moral experience by imposing an artificial taxonomic framework on it. This approach seems to imply that that subsequent prioritization will take care of itself, or that it is a far less important problem than classification.

There is another sense of propertarian that is just as important as the first. It refers to the political economic dimension in which philosophical and practical ethics as such is itself embedded—as is any other human activity. In our dualistic tradition, both animals and things are readily categorized as *property* while human beings usually are not (at least not in recent times). The dualism person-property is another of the hegemonic binaries of the Modern constitution, and is built around distinctively western conceptions of both persons and the institution of (private) property. This insight is important in particular for environmental ethics because the institution of property relations conditions virtually all of our behavior and decision-making regarding nonhuman animate and inanimate beings, including collective entities such as woodlands, mountains, seas, and even the atmosphere. In the current era, as Rousseau noted long ago, fencing in (enclosing) parts of the Earth magically transforms that portion of land into "private property," the owner of which is, under the right conditions, within her or his rights even to kill "persons" in order to preserve it. This valuation of property over person is just one extreme example of disordered priorities. In other words, "property" is not simply a matter of politics while "person" is a matter of ethics; they are defined in relation to each other. On the dualistic view, persons can be property holders, but they cannot be property; animals and the land can be property, but cannot be persons. The relation of "possession" is what connects the two conceptions of propertarianism introduced here. This has its ontological basis in substance metaphysics. In ontology, the substance-property dualism entails the relation of possession; substances possess properties. Likewise, social property assumes the legal subject first exists and then something belongs to it, making the relation of possession secondary to the subject. Among other things, keeping this connection between "having properties" and "owning property" in view allows us to take seriously the developmental context of consumerist culture and property relations in which an increasing number of human beings have been reared. In this context, property itself should be regarded as an ontologically basic metaethical condition of ethical relations. Thus, for propertarianism, we first want to know *what properties it has* or *whose property it is* in order to determine our moral responses to entities theoretically and practically.

To take the problem of prioritizing values in cases of conflict to be ethically primary differs fundamentally from propertarianism. Interestingly, this distinction was not unknown to earlier writers, but for the most part they set aside such clashes and conflicts because they seemed to believe that answers to the question "which beings are morally considerable?" (i.e., which properties confer moral relevance or intrinsic value) were logically prior to answers to the question "whose interests take priority in cases of conflict?"[18] In perhaps the most influential and often-anthologized essay "On Being Morally Considerable" (1978), Kenneth Goodpaster made a clear distinction between "moral considerability" and "moral significance" relevant for ethical theory broadly, but as he saw it especially relevant for environmental ethics.[19] Where moral considerability concerns the question whether an entity falls into the moral ballpark, "moral significance" pertains to "*comparative* judgments of moral 'weight' in cases of conflict."[20] Where the former concerns "necessary and sufficient conditions that should regulate moral consideration," the latter includes "questions dealing with how to balance competing claims to consideration in a world in which such competing claims seem pervasive."[21] Finally, he hinted that answers to the first type of question will not lead directly to answers to the second, since the criteria for judging them differ: "We should not expect that the criterion for having 'moral standing' at all will be the same as the criterion for adjudicating competing claims to priority among beings that merit that standing."[22] This is a perceptive and well-expressed distinction. In focusing exclusively on considerability in that essay and setting aside the issue of significance, however, we might say that Goodpaster set the stage for what followed in environmental ethics for at least two decades. Goodpaster was just one of several writers who at the time recognized the importance of such a distinction.[23] Good logic seemed to demand that the taxonomic issue be settled first, even if many recognized the significance of "significance."

What I am claiming is that propertarian approaches draw taxonomic boundaries around the moral community by determining which entities are worthy of our moral regard, or define the minimum threshold for entry into the moral universe. Propertarian approaches give the impression that metaethical epistemological and ontological issues are the sine qua non of environmental ethics.[24] These differ markedly from "priority" concerns, since—in an already value-abundant world—they bear on competing "significances," entailing reasons, values, or motives for acting which express preferences, prioritizing, and ranking of values by the moral agent in community. I argue that questions of prioritization are the primary or central dilemmas in our ethical lives, but given the Modernist theoretical legacy, consumerist social imaginary, and the political economic institutions we inhabit, propertarian questions have mistakenly been taken to be primary. The question is not whether something has value at all—since our most minimal dealings with anything tacitly affirm that it has some value—but whether it has more or less value relative to the value of something else. For instance, judging by their behavior, many Americans do not think the value of a cow's life outweighs the human pleasure of eating its flesh.[25] It is safe to say that while the distinction between propertarian and priority issues has been observed by a number of authors, priority issues—when treated at all—have nowhere been treated as extensively as propertarian issues. It is high time to address this shortcoming in environmental value theory.

Priority questions take value conflicts as point of departure and reflect on how some values may be prioritized over others by agents individually or communally. The experience of conflict and competing claims is all-pervasive, and most environmental issues are readily articulated and grasped as matters of competing claims to priority. For example, is draining a wetland for real-estate "development" more important than maintaining the ecosystem services and biodiverse population it contains? Is a comfortable lifestyle for a few people and profit for a few developers more important than its vital life-support services for humans and nonhumans alike? Is clear-cutting a forest for lumber, fuel, or pasture for livestock more important than using it as a carbon sink for greenhouse gases, or letting it remain wild? Are monoculture plantations of eucalyptus more important than sustainable livelihoods for people and native ecosystems for bioregions? In a now classic North American case: is the existence of the grey-horned owls in Northwest forests more or less important than the livelihood of the loggers laboring in them? Even these moderately complex cases show that questions of priorities have a primacy in human experience that is poorly captured by the preoccupation with propertarian concerns. They show that environmentalists need a richer values discourse of qualitative contrasts in order to articulate these competing priorities. According to the perspective introduced in the last chapter, moral agents daily grope for languages capable of facilitating their ability to qualitatively contrast the various interests (needs, values, claims, demands) at stake in moral conflicts, and such groping articulations help individuals and communities approach action with a clearer vision of priorities. Such languages are constitutive of our individual and communal identities, and entail strong evaluations that implicitly prioritize values. In addition, our personal narratives implicitly or explicitly refer to tacit priority rankings of values that structure the cultural and individual identity-forming ethos. Assertions regarding values are always contextually qualified. "I love animals (but not enough to die for them, to protest at a CAFO, or even to stop eating them)"; "I hate capitalism (but not enough to give up my job and be an activist, join the local Occupy group, or cut up my credit cards)." The most important question is not whether things have value-conferring properties, but the question about clashing value contents themselves. If part of the mission of environmental ethics has always been to motivate or guide (environmentalist) action, then a value ethics which explicates and reflectively engages self and community with diverse priority schemes is an important part of this mission.

This turn to priorities requires adoption of some "enabling environmental practices" that deliberately set aside propertarian questions, and these include narrative approaches, a virtue-ethical interest in the stance of the agent, and communicative ethics that generously presume the agency of those beings with which we are in constant interaction. Since propertarianism in ethics is an expression of its social context, we also need to remain critically aware of the ways that this context shapes ethical theorizing. It is against the hegemony of this context that enabling practices take shape. We have to reject the notion that establishing universal moral minima is our highest goal, and instead foster ethical creativity in a world of competing priorities. In the next section, I will outline the concept of enabling environmental practices and other prerequisites for making priority questions as primary in theory as they are in experience.

ENABLING ENVIRONMENTAL PRACTICES AND VALUE THEORY

In the ontogenetic life context of growth, reproduction, and survival, practices have more ethical traction than do simple conversations about good and bad behavior, but since we are self-interpreting animals, articulating what is valued in languages of qualitative contrasts also has the potential to reflexively reshape the practices themselves. Crude distinctions between *human* and *animal, person* and *property* provide relief from overstimulation by freeing those who employ these terms from the burden of moral concern in each and every interaction with each and every being they encounter. Children simply imitate their caregivers in this respect, learning not only what can be done with others (intentional affordances) but what should be done with them (normative affordances), and how much caring labor to "invest" in relationships that matter. In a social context and ethos where nonhumans and even other humans are regarded as means to be exploited for human ends, a pattern of instrumentalization is cultivated, overpowering more nuanced and caring modes of interaction with crude stereotyped responses. Dualistic hierarchies have their full force here. If we are to avoid indulging even the benign forms that propertarianism takes in ethical theory, then we have to cultivate what Plumwood called "counter-hegemonic" practices of dehomogenizing, rendering difference nonhierarchical, foregrounding rather than backgrounding relations of dependence, and undermining not just instrumentalist action, but the dualistic structure on which instrumentalism is built. Enabling environmental practices refer both to conceptual shifts as well as to everyday practices that put into question propertarian habits and strictures, and smooth the path toward fully appreciating the primacy of prioritization.

In the 1980s and 1990s, Anthony Weston and other pragmatists criticized intrinsic value theory on a number of grounds. Not only was it conceptually flawed in its quest for self-sufficient, abstract, and specially justified values, but more damagingly, this quest was seen to be just one more manifestation of broader cultural tendencies toward the instrumentalist reduction and simplification (or "disenchantment") of the world. Far from being a way out of this situation, according to Weston "the appeal to intrinsic values further compartmentalizes an already fragmented world."[26] Plumwood too noted that any new conception of self and ethics has to contend with the institutional dominance of instrumentalism in the public sphere. She claimed that "ecological selfhood cannot be conceived in terms of the thunderclap of personal conversion to an after-hours religion of earth worship, tacked on to a basically market-orientated conception of social and economic life." She continued,

> It must be seen rather as an attempt to obtain a new human and a new social identity in relation to nature which challenges this dominant instrumental conception, and its associated social relations. Hence it is a *practice of opposition* which parallels that of the attempt to retain and expand other non-instrumental forms of social and economic life in the face of relentless instrumentalizing pressure.[27]

Weston also argued that environmentalists should be in the business of presenting "open-ended challenges" to the status quo, "to create the social, psychological and phenomenological preconditions—the conceptual, experiential or even quite literal 'space'—for new or stronger environmental values to evolve. Because such creation will 'enable' these values," he calls this "practical project *enabling environmental practice*."[28] According to Weston and Cheney, fully embracing the conception of "enabling environmental practices" in the theoretical register also means rejecting the standard model of the relation between epistemology and ethics. The standard model holds that ethical action is a response to our knowledge of the world (knowledge comes first, then ethics follows); the world is readily knowable (what we need to know to act ethically is completely given); ethics is inherently incremental and extensionist (we begin with the human ethics we know, and expand from there); and the task of ethics is to sort the world ethically (the propertarian problem of moral taxonomy). Ethics conceived at least partly as enabling environmental practices entails accepting a different set of assumptions: ethical action is an attempt to open up rather than foreclose possibilities; hidden possibilities surround us at all times (the world is not readily knowable); ethics is pluralistic and discontinuous (not smoothly extensionist); the task of ethics is to explore and enrich the world (working against ready-made classification).[29] This critique of existing metaethical assumptions resonates strongly with my own, and other elements of their perspective are valuable for preparing the way for dealing with priority questions. The concept of "enabling environmental practice" allows us to both remain in contact with the practices and institutions that sustain current anthropocentric culture and to experimentally intervene with nonanthropocentric alternatives.

A powerful example of the role of discourse and social learning in structuring behavioral expectations, echoing the discussion of circular processes in a previous chapter, is what Weston called "self-validating reduction." Self-validating reduction provides an illustration of the way that categories, valuations, and practices can be mutually reinforcing. Cows imprisoned in concentrated animal feeding operations (CAFOs) are mechanically fed in their tiny cages or stalls and have been bred to be docile, and their physical constraints also induce submission. Then they are claimed to be sluggish, unresponsive "brutes." Dark-skinned people are stereotyped as unintelligent and only fit for manual labor while at the same time they are socially deprived of opportunity and education. Women are essentialized as child bearers and still in many societies socially discouraged from doing anything else. We can think of these processes as more elaborate, cultural forms of the circular processes described in a previous chapter, through which humans perceive and communicatively interact with the world in order to discover what things can do. For social, self-interpreting animals seeking stability and regularity, it is a small step from considering what they can do (their affordances) to unilaterally dictating what they should do (normative affordances). Conceptual and practical stereotypes inculcated in children by the adults who surround them provide comfort and relief by reducing complexity, and building institutions around these stereotypes allows agents to save effort by having a whole range of decisions ready-made for them. Without such reduction or relief, we would always be in the situation where the only limits to our responses to the world and others would be our energetic and understanding limits of care—we could potentially care for everything and everyone. In such a situation, we apparently feel the need to limit the class of things about which we can care, so we both describe them and treat them in a limited way. One of our enabling environmental practices must be to interrupt what Weston calls this self-validating reduction by breaking these cycles of reinforcement as a way of fostering cultural change. Seen in this light, a question about the rate of such change arises.

A naturalistic approach to such dilemmas entails accepting that cultural change happens at varying rates. Weston thinks—like Bender, Plumwood, Gare, and others—that what environmentalists are engaged in is a movement to create sweeping cultural change, not simply changes in narrow ethical conceptions. If our contemporary anthropocentric ethics are themselves expressions of prevailing beliefs and attitudes that have resulted from many cultural coevolutions of values and practices, then any changes in them will likewise be the results of slow processes of change. This insight demands that we look at the ways that values "are deeply embedded in and co-evolved with social institutions and practices."[30]

> Fundamental change (at least constructive, non-catastrophic change) is likely to be slow. Practices, habits, institutions, arts, and ideas all must evolve in some coordinated way.... Thus it may not even be that visionary ethical ideas ... are impossible at any given cultural stage, but rather that such ideas simply cannot be recognized or understood, given all of the practices, experiences, etc. alongside of which they have to be placed, and given the fact that they cannot be immediately applied in ways that will contribute to their development and improvement.[31]

Even "radical social criticism" draws on "the historical and social embeddedness and evolution of ideas" to retrieve forgotten or suppressed cultural values. As we have seen, a liberation model of anthropocentrism, for example, is rooted in such contexts. Creating narratives that articulate the continuity between the current struggle and past struggles or values in the very same tradition is important for maintaining a sense of stability beneath the dynamism of changing priorities. Long periods of experimentation with coevolving values and practices are to be expected, as well as uncertainty about eventual outcomes. Weston argues that if environmental ethics is at just such an exploratory, early stage of development, then we can only barely glimpse what a nonanthropocentric ethics and culture would look like. This suggests that the task of environmental ethics is "to *open* questions, not to settle them," and we must consider all attempts as "open-ended sorts of challenges."[32] With reference to the anthropological model proposed in the previous chapter, a few sorts of enabling environmental practice can foster this work, including narrative redescription, dialogical communicative ethics, and attention to dispositions or virtues of moral agents.

In our local-scale reproductive and developmental practices we can work to redescribe moral-environmental situations in noninstrumentalizing and nonanthropocentric ways. This practice should be considered at two temporal-spatial scales: at the ontogenetic scale of the situations in which values come into conflict and where priorities are ranked, and at the historical scale of long-term cultural change. In a context where there is value blindness or impoverished perception, a large part of the ethical task is simply to qualitatively describe the case or situation at hand where the values conflict. Specific conflicts will need to be articulated by participants in their community context, and even this process itself can help to reveal undiscovered values obscured by instrumentalism, and to revitalize the affective appeal of values that tend to be crowded out by institutional and dispositional patterns. Just doing the articulating in new narrative forms or conversations goes a long way toward allowing people to see and feel underappreciated values, and to consider which values ought to take priority over others.[33] Tomasello said that language itself contains the imaginative variation such situations may demand (as explained in the last chapter). If the same plant can be described as a weed, invasive plant, or flower; if the same ecological pattern can be described as a stochastic fluctuation, ecosystemic regularity, or superorganism, the evaluations embedded in these descriptions can be made explicit and assessed. Articulations, as forms of enabling environmental practice, may take the shape of stories and narratives. "Stories provide a more nuanced, 'ecological' understanding of our place in the world—including our *ethical* place. Stories are the real homes of so-called thick moral concepts, concepts in which evaluation and description are so intertwined as to be conceptually inseparable."[34] O'Neill, Holland, and Light also emphasize the role of narrative and thick description. They claim that we must

> begin ethical reflection from the actual thick and plural ethical vocabularies which our everyday encounters with both human and non-human worlds evoke. If we start with a thick and plural ethical vocabulary we invoke a similarly thick and plural set of relations and responses appropriate to different kinds of beings. These are lost if we start from a picture of moral theory as an exercise in the derivation of specific moral norms from some set of moral primitive concepts or propositions.[35]

The importance of narrativity in reflecting on ethical tensions has also of course been highlighted by feminists. The narrative model claims that we should not abstract from the situations we enter into in order to assign intrinsic values and rights, but "should enrich our description of the situation as much as possible, enrich it, in fact, to the point that appropriate care simply *emerges*."[36] The very articulation of the situation in terms of varying qualities of significance can give us clues about the moral course of action, without reducing these qualities to the abstractions of rights, universal rules, or principles. Aiming at an intermediate level of description, that is, intermediate between too much abstraction (which provides quick but empty relief) and too much specificity (which loses us in the details of the case and is overwhelming), allows us to contextualize and interpret persons and situations as exemplary samples of value qualities. For example, imagine an aged farmer struggling to look after himself on an isolated farm visited by a social worker. She suggests he move to a retirement home. On an objective list account of his well-being, this would seem to be a sensible choice. All of his life conditions might improve. But what is missing is any consideration of the farm he has worked for sixty years, its having been passed down through his family for generations. To stay rather than leave is appropriately to continue the developmental narrative of his life in that place rather than radically disrupt it by abandoning his home. It is "more important for him to stay" than to leave and live elsewhere. The implicit structure of priorities embedded in the narrative should be noted. "Preserving relationship" with a place, loyalty to tradition or family, felt kinship with the land, and courage to go on independently may be some of the moral values involved in the account. Consideration of such articulations as a necessary methodological aspect of an enabling stance, as I have suggested, follows directly from the ecological materialist, developmental conception of the human.

Although it is an indispensable enabling part of the priorities approach, however, narrative is not sufficiently robust in itself to guide us through the thicket of conflicting values manifest in environmental dilemmas. While such an approach allows us to begin to bridge the gap between evaluation and action, it also has its limitations. The narrative approach itself does not guarantee that individuals in communities will not just continue the same capitalist or consumerist narratives of growth and technological triumph that have led to so much ecological devastation. Narrativity has been opposed to proceduralism and rationalism, but not all narratives are good just because they are narratives. As noted in the previous chapter, there can be disagreements about articulations, and there is no single best account. We have to recognize the fallibility of our articulations and enlist the judgment of others. To be enabling, transformative or counterhegemonic, narratives should be informed by multiple voices, encompass whole ranges of contrasting values, and be structured by the historical and developmental scale of reproductivity and flourishing for particular groups or communities. As I will argue in the following chapters, a conversation about conditioning and conditioned values and their prioritization lends clarity to the articulation of motives and environmental values that may be initially vague and inchoate, and brings some order to their relations. Values discourse adds an indispensable dimension to narrativity. If the narrative gives you an indication about how the story is to end, the relations among values themselves provide a second-order logic of priorities. Without the latter, insular, oppressive community narratives can continue just as much as liberatory ones may be initiated. In addition, to dwell exclusively on narrative articulations carries with it the risk of remaining overly anthropocentered, since the focus may remain more on human experience than on the more-than-human world.

Plumwood makes it clear that we have to adopt many strategies to resist the dominant conceptions. In order to avoid the potential anthropocentrism of narrativity, she argues that we must also embrace a "communicative ethics."[37] If "our possibilities for interaction with and perception of the world are influenced in major ways by the postures we ourselves choose to adopt,"[38] then we need to be open to the other as a communicative being, and this virtue may be expressed in our chosen stance. Plumwood recommends adopting the "intentional recognition stance," which is "one of a number of counter-hegemonic practices of openness and recognition able to make us aware of agentic and dialogical potentialities of earth others that are closed off to us in the reductive model that strips intentional qualities from out of nature and hands them back to us as 'our projections.'"[39] Communicative ethics entails getting away from the dominant intellectualist conception of communication, which has predominated at least since Descartes declared that animals do not possess language or thought.[40] This is an essential prerequisite for taking Other-agency seriously.

> Rationalist models which treat communication in intellectualist terms as an exercise in pure, abstract, neutral and universal reason, and which delegitimate the more emotional and bodily forms and aspects of communication, operate to exclude non-humans from full communicative status just as they exclude various human others accorded lower human status as further from the rational ideal. These disembodied rationalist models exclude the forms of communication associated with animals along with

the forms of communication associated with women, with non-western cultures and with less "educated" classes.[41]

This entails, among other things, a massive expansion of the concept of communication and communicability. It means spreading

> the category of potential communicants and the concept of communication and communicability out very widely beyond the human to take in not only living inhabitants of the earth and of space, but also places, experiences, processes, encounters, projects, virtues, situations, methodologies and forms of life. In a dialogical methodology, the other is always encountered as a potentially communicative other. This is part of what is involved when we move from the reductive subject-object models of relationship characteristic of mechanism to subject-subject models of an alternative communicative paradigm.[42]

Much of human communication is bodily, mimetic, gestural, and expressive, as is animal communication.[43] We often sense much more about what nonhumans (such as our pets) want in this way than we let on, since defining communication narrowly allows us to systematically ignore this information in order to insist that there is nothing forthcoming to be taken into account (facilitating propertarian categorization and relief).[44] Plumwood also recognizes that adopting a communicative ethic is not a magic bullet, since communicability "does not automatically eliminate the dynamic of power."[45] The communicative ethic does get us further into the realm of inter-Other dialogue, but it is not a cure-all. It is just one among other enabling environmental practices.[46] It is an important one, since if we are to "treat at least the general goals of the other's well-being, ends or *telos* as among our own primary ends," then those others have to be regarded as communicative agents in their own right, that is, as having their own ends that can be furthered or frustrated. This entails a revision of dominant ontological frameworks as well, since in the Modern constitution nonhumans have been considered nonagents. I consider this point further in the last chapter of the book.

Finally, attention to cultivation of virtues by the embodied and embedded moral agent is another sort of enabling practice in addition to articulation and a communicative ethical stance. This is because virtue ethics generally also resists theoretically dualistic construals of human motivation, and emphasizes relational selves. Characteristic of an ecofeminist ethic, along with rejection of all forms of oppression (including speciesism), is contextualism and pluralism, and placing the virtues (or values) of care, love, friendship, appropriate trust, reciprocity, kinship, and sharing at the basis of ethical concern.[47] According to Plumwood, feminist ethics recognizes the central role of "special relationships," such as family ties, with which our moral education begins. The social-natural context of growth, reproduction, and survival is one of special kin relations long before it is one of "human rights." In relation to environmental ethics, this means that "with nature, as with the human sphere, the capacity to care, to experience sympathy, understanding, and sensitivity to the situation and fate of particular others, and to take responsibility for others is an index of our moral being."[48] Plumwood considers the type of ethics developed in feminist circles to be a kind of embodied virtue ethics.[49] The virtues of "friendship, love, respect, care, concern, gratitude, community and compassion" resist the reason-emotion dualism, and are often in conflict with the rational instrumentalism of the public sphere and can be strategically employed to intervene in this sphere. Such accounts do not dismiss the claims that the "other" who is the implicit correlate of these concepts be treated as "intrinsically valuable" in its own right, and this is even a necessary "theoretical complement of a virtue account of ecological selfhood."[50] In addition to those virtues often listed, Plumwood refers to others that have emerged in the debates, including openness to the other, generosity, ability to put oneself in the place of the other and respond to their needs, sympathy, and "recognition of specific relations of dependency, responsibility, continuity, and interconnection, as well as those of difference (including human difference) and of respect for the independence and boundlessness of the other."[51] In the next chapter, I will talk more about the differences between virtues, moral values, and goods values, but here you can see that virtues like generosity, sympathy, and care bear on the right kind of relation to goods or others who are already assumed to be of independent value.[52]

All of these strategies are central preparatory contributions to a viable environmental value theory, and must be preserved and promoted. However, even with these stances in place we remain at the threshold of the messy, value-saturated jungle of environmental value conflicts. A pluralistic value theory should allow us to structure consideration of values and value commitments in order to more effectively communicate the significance of value priorities in situations of common concern. If we cannot even describe conflicts in such a discourse of values, then we have no hope of articulating a future in terms of new configurations of value priorities either. The embodied-embedded, ecological, acting self-in-relation is the starting point for a value theory for environmentalism that begins by undermining the formalism of propertarian approaches and questions, since the epistemology-ethics relation they adopt is fundamentally flawed. A value ethics that begins with the experiential richness of the world, adopts a narrative and communicative stance, and a metaethics of an embodied and embedded self is better equipped to explore the domain of value prioritization than other existing approaches. The following section summarizes some of the historical and theoretical context of the approach to value theory taken here. In the next chapter I will consider whether, among the diversity of values and different cultural and individual ways of prioritizing them, meaningful agreement can be reached about a ranking of values that might often place environmental values above at least some human interests. On the view that begins in a world of abundant, competing values and agents who must choose in face of them, there are no absolutes, but there are real and axiological relations of dependence. Environmentalist prioritization takes place in recognizing dependencies that preserve stability and variability of patterned valuation in growth, reproductive, and developmental contexts, and in community and communication with human and more-than-human others.

On Value Theory

What I have been calling *value theory* or *axiological ethics* here differs from the now conventional triumvirate of ethical approaches presented in ethics textbooks—utilitarianism, Kantian deontology, and Aristotelian virtue ethics (all of which have been tried by environmental ethicists)—by highlighting the perception and prioritization of values in its consideration of moral normativity. There are both Anglo and Continental traditions in axiological ethics, including figures such as G. E. Moore and W. D. Ross in England; John Dewey, Ralph Barton Perry, and Stephen Pepper in the US; and Franz Brentano, Alexius Meinong, Max Scheler, Dietrich von Hildebrand, and Nicolai Hartmann in Austria and Germany. Most recently, philosophers Charles Taylor and Joseph Raz have represented aspects of this value ethical tradition. The value theories in this affinity group certainly do not represent a single school of thought, but do share some similarities. Despite their many differences, one element they have in common is the attempt to consider what is valuable or worthwhile in human experience generally (and usually, but not exclusively, in relation to moral agency).[53]

The value theories I draw on here share several other characteristics that follow from their account of moral agency and experience. They agree that the human experience of value is inherently plural, and that values at least sometimes conflict, apparently irreconcilably. Many values occur in clusters of related values, and values are items about which disagreement seems inevitable. One feels obligated to choose among them, and they are experienced as higher or lower, or as more central or peripheral. Finally, phenomenologically speaking, we spontaneously relate to values as nonsubjective (not as self-produced or projected).[54] The values perceived are taken to be there in objects, persons, or situations, and are initially not regarded as fictions or illusions.

This smattering of insights leads to a rich array of consequences, more of which will be discussed in subsequent chapters. A few are outlined here. We often feel that the existence of some values "offers the same absolute resistance to the will of the subject as any real object of perception," and even that at least some of these values are objective.[55] On this account, *objectivity* means that values have a "constitutional relationality" to embodied, ecological, moral agents such as ourselves. What is ontologically or constitutionally relative, but nevertheless invariable about them, is that values are the kinds of things that are always "for someone" (human or nonhuman agency), in precisely the same invariable way that geometrical categories apply to geometrical figures qua geometrical, and biological categories apply to organisms qua organisms. Their existence is "relative" to beings who evaluate.[56] Moral agents possess a primary, affective-conative-cognitive perception of values which may be narrow or wide, coarse or refined, depending on degree of cultivation. As I explain further in the next chapter, values are correlates of affective intentional acts such as wonder, awe, love, respect, and veneration, and disvalues are correlates of disgust, hate, and other negative affects. Our perceptions and affections give fast-system order to our value preferences and avoidances, and disclose the more or less rich domain of values to our value sensibility. Pragmatically, this sense can be measured by the richness of linguistic expressions for values that one uses. Given the nature of circular processes in social learning, the more value terms one employs to describe an experience, the more likely one will be to see and experience such values in the world; the fewer, the more truncated the view. Languages of qualitative contrast help us to map the terrain of natural-social space, orient ourselves within situations, and gauge our own moral responses to them. This is not a matter of direct internal self-evidence (intuitionism), but of language as an extended perceptual organ that reveals qualities of ourselves, others, situations, and relationships. For every human being striving to orient themselves vitally and morally in a cultural context, values have an "absolute" sort of existence. That is to say, the most important of them are experienced as *independent* of our will, and we would not be ourselves without (reference to) them. Some things self-evidently have value or "ought to be" over which we have no prerogative—the sun, soil, atmosphere, and water are valuable to more-than-human beings as well as ourselves. In other words, the qualitative descriptions of various communities may converge on the significance of particular practices, beings, or relations that are valuable or damaging to both people and ecosystems beyond the clumsy opposition between anthropocentric and ecocentric ethics. To compare divergent value rankings we need not refer to a supposed ontologically detached realm of values, only to articulations of those values that presuppose the existence of an orienting rank order embedded in the constitutional relations between valuing beings and the existing or potential real bearers of those values. That there is such an important domain of value which plays a significant role in our self-conceptions and ethical decision-making has been the theme of much contemporary moral philosophy. What these values are and whether in cases of conflict they can be prioritized has been given much less attention. I will try to remedy this in subsequent chapters.

CHAPTER FOUR

Environmental Values and Vital Priorities

ENVIRONMENTAL GOODS AND MORAL VALUES: BASIC PRINCIPLES

The ecological materialist anthropology sketched in the previous chapter considers the human situation in the world to be one of experienced conflicts of values that we are compelled to prioritize in order to orient action, reflect on it, and justify it. The human plasticity of impulses gives rise to an enormously wide range of valuations that are harnessed and stabilized through socialization and social institutions. The quest for intrinsic value in nature can be regarded as a symptom of the typical human desire for stability and relief amidst a buzzing world that overburdens us with stimulation and provides scant guidance about how to respond to it. The need for relief is a result of our "unfinished" nature. This indeterminacy of the human condition is usually mitigated by embracing habitual modes of behavior and institutional channeling of impulses that simply reinforce the reigning cultural priorities. These implications of the metaethical discussion can be formulated in four general principles for a pluralistic value theory.

The first principle of valuing for a being burdened with an "excess of impulses" is that (1) *anything can have any value*. I will call this the *relativist principle*, whose importance is capitalized on by advertisers but often overestimated by constructivists and subjectivists. However, this relativist principle is necessarily constrained by the real world, since (2) *not every situation, thing, or being can bear just any value,* given both its constitution and its relations to the kind of being we are. Let's call this *Rolston's principle,* since he is the philosopher who insisted most strongly on the mind independence of value bearers. Thirdly, imitative learning, instrumental parenting, and other modes of teaching that make up our cultural traditions and social and environmental contexts provide "guardrails" for evaluation by (3) *precognitively or affectively "loading" things and relations with certain values (and meanings) but not others.* The total context guides us in selecting which of these values is treated as "ultimate." We could call this the *ontogeny principle,* since how you were raised to appreciate the values in things is a key condition for your current valuations. Finally, this plastic but conditioned precognitive loading with values (4) has to be regarded in terms of the *reproductive and developmental life span of individuals and cultures.* Even if a bearer is fit to carry a whimsical or arbitrary value, it normally cannot do so *indefinitely* in a survival context. This is the *survival principle,* where survival means cultural-biological flourishing. These principles also imply that there are no unequivocal anchors of value "out there" that will determine individuals in community to behave in particular ways, but that there are innumerable perfectly real value qualities borne by natural and socialized entities that may serve to stabilize human impulses. Principles 2 and 3 already show how the first relativity principle is constrained, and principle 4 might be regarded as containing the previous three. Principle 4 is both substantive and pluralistic: it does not say what values there are, but it does regard the material ecological condition of human existence as a baseline. It entails that where what I will call *vital values* are systematically violated, the conditions for social and ecological well-being are undermined.[1] I will explain exactly why this is the case in the last section of the chapter.

An example might help to illustrate: (1) In the context of preindustrial cultures, petroleum may have no value, or it might be regarded as the Earth's blood in story and song, or used aesthetically as body paint. In industrialized cultures such as ours, it comes to occupy a privileged place, fueling much of the human activity of civilization, and to keep the oil flowing nations are willing to go to war (relativist principle); (2) however, oil can under no circumstances bear the value of nourishment for beings like ourselves, and is in fact toxic to most life (but not to oil-eating bacteria). It can be refined into gasoline, diesel, kerosene, and plastics, but it cannot have just any value qualities whatever (Rolston's principle); (3) in current petrocultures, the economic system has become so dependent on the substance that every child implicitly understands its value from an early age; it is like water or air, something that keeps society churning along. It is preloaded with value in all of the networks of relations in which it figures (ontogeny principle). Finally, as those in positions of power are too late to realize, (4) the era of cheap energy cannot continue indefinitely, not just because the reserves will one day run out, but because the more the stuff is burned the more the atmosphere is burdened with its components, imperiling the long-term cultural-biological survival of humans and nonhumans. Petroleum is thus at best an equivocal "good" on the back of which many outright brilliant and many questionable cultural achievements have been erected, and a "bad" that produces ecological degradation and human suffering in the long term (survival principle).[2] The example simply shows that not only must certain kinds of real bearers exist in order for certain kinds of values to exist, but also that some kinds of values are directly dependent on the existence of other kinds of values. Rather than continue to speak about "values" in this general way then, it is time to define the various classes of values and their relations. Exploring their relations and dimensions for the purposes of environmental value theory is the task of this chapter. An exhaustive discussion of pluralist value theory cannot be provided here.

TWO KINDS OF VALUES

When we consider the human as an acting being, that is, one that acts purposively in natural-cultural places and orients itself in a world not of its own making, a two-tiered taxonomy of values initially emerges. We may distinguish between the *value of the ends intended in action* and the *value of the intention itself*, a distinction which follows naturally from the structure of action. For instance, the eco-sabotage activist pours corn syrup and sand into the gas tank of the bulldozer in order to prevent it from clearing the forest; the animal rights protesters march with signs in front of the labs where cosmetics are tested on animals in order to generate awareness of their plight and hopefully prevent further nonhuman suffering; the Zapatistas in Mexico march silently by the hundreds "in order to be heard." In each case, agents intend to realize some goods as the ends of action, although these goods values are themselves not moral values. A clear distinction is evident between the good aimed at (forest preservation, continued animal life free of suffering, and eco-political liberation, respectively) and the value of the agent's intention itself, which is a moral value that may be described as noble, daring, courageous, compassionate, heroic, righteous, or free. These latter values are not intended in the action, but "ride on the back of" the act and are realized in social-natural contexts of action.³ A value theory that fails to recognize the difference between values intended (goods values) and the value of the intention (moral values)—such as intrinsic value theory, or economism—cannot even begin to capture the most significant moral dimensions of environmental evaluation. On my view, it is obvious that the intrinsic value something might possess is not itself a moral value, even if it might be regarded as a good that could evoke moral responses from moral agents. This is even more obvious in the case of economism. This also means that there is nothing inherently moral about value as such. There are moral and nonmoral values, and the world is full of them. This relation in the structure of action gives rise to the taxonomic distinction between the classes of *goods values* and *moral values* that I will employ here.

Goods values can be means or ends, occur in experiential clusters, and can be carried by things, organisms, systems, processes, situations, persons, and any other item that counts in a cultural or ontological taxonomy. The relativist principle is in evidence here. The aims of environmental action have included preservation, conservation, or restoration of nonhuman life, wildness, integrity, environmental justice, "goods of their own" of individual nonhuman lives, nature's fecundity, and other qualities. Of whatever sort, such goods are things taken to be real and "constitutionally related" to the vital capacities of living beings.⁴ They are "good for" a being of a certain kind, in an objective sense of "for."⁵ Moreover, goods values are constitutionally relative to humankind as both living and cultural beings. Many goods are only goods in the eyes of persons, discovered in symbolic, cultural perception.⁶ The (goods) value of high achievement test scores is only evident to humans, and moreover, humans in a very specific cultural context. The moral worth of the activist resisting corporate palm oil monoculture in the Borneo is not appreciated as such by the orangutans for whom they struggle. While environmental goods have often been the focus of the discussion of environmental values, the more recent turn to virtue ethics, moral psychology, and narrativity (thick description) in recent ethical theory also draws attention to the motivations and character of moral agents-in-community.

Moral values, which might include virtues like courage and self-control, ideals like brotherly love and charity, Enlightenment values like freedom and fairness, or even the Nietzschean or anarchist value of self-creation or self-realization, are values which can be borne only by moral agents. They can only belong to creatures with the ability for "second-order valuation," that is, those not only able to be oriented by desired goods values, but also able to ask *whether this or that orientation to these goods is itself a good orientation to have.*

On this view, wombats and fishes, while they are considered value-seeking or goal-directed *agents* with goods of their own, are not *moral* agents.[7] While this may seem to run the risk of preserving the traditional human-nature dualism, we have to keep in mind Plumwood's dehomogenization strategies and nonhierarchical concept of difference. According to Plumwood, living things should be approached as communicative others in a dialogical stance of openness, making room for every sort of agency expressed by them. Groundhogs are vital agents who burrow and sometimes threaten foundations, but they do not seem to exhibit behavior which would lead us to attribute to them the desire to be a better groundhog. This stance grants that the groundhog has its own vital value independently of our evaluation (good of its own), and our intention adds another layer of (recognized-articulated) valuation to its (mutely enacted) conative good of its own.[8] Not counting it as a moral agent in its own right does not take anything away from the groundhog, since—based on the arguments of the previous chapters and contrary to the dominant propertarian assumption—its possession of certain properties is no longer the decisive consideration that determines how we should respond to it. In order to avoid propertarianism we should remember that what a being is does not provide a limit on our moral regard for it, as if the base class of moral consideration has a frontier to be discovered. We can morally consider all things, but not all beings in the world morally consider.[9] That is the burden and promise of humankind.

NOTE ON THE ONTOLOGY OF VALUE

Before proceeding, let me try to remove one likely obstacle in the way of accepting a theory of values of this sort. Many writers avoid discussion of values because the term can evoke bad memories of Platonizing metaphysics and an ethical intuitionism that simply tells us directly what is valuable. For better or worse, however, value terms cannot be avoided and are indispensable for discussing ethical issues and for articulating a common future. Fortunately, there are other ways of thinking about the existence of values. In this book, the "ought" of moral values is empty without reference to concrete value qualities, and these qualities are *exemplified* in social acts.

With reference to both goods and moral values, recall the concept of constitutional relationality from the previous chapter. Values have a constitutional relationality to embodied, ecological, moral agents such as ourselves. What is ontologically or constitutionally relative, but nevertheless invariable about them, is that values are the kinds of things that are always "for someone" (a human or even nonhuman agent). As mentioned above, moral values are qualities of social acts and persons. Once generated in these acts, they no longer depend on individual subjects and play their role as detached normative criteria for judgment. They are cultural products that take on a life of their own and are directly *inscribed* in our affective social lives (at minimum). Values are names for complexes that include social acts or practices, dispositions, affects, and normative "oughts." To refer to an earlier example, a teacher hands out the same textbook to each student at the start of the term, with the intention of fairness, and could be judged by the children or by herself as practicing a fair distribution. This is a *sample* of fairness that can become *exemplary* for future acts and normative judgments. Values have been mistakenly regarded as possessing the same kind of being as ideal objects such as mathematical formulas because they easily become reified in language. But value terms are simply convenient names for such social complexes built around a sample act. The concept of the exemplarity of values allows us to set aside the Platonizing question about the ontology of values for now. Exemplarity designates the way in which a case can serve as an example from which to derive a rule, or an "ought."[10] Samples are used as principles of imitation for social and moral behavior. This way we are not misled into thinking that we need an abstract idea of justice or freedom in order to recognize just or free acts. The value as such *can* be "detached," compared and contrasted with others, placed in a rank order, and so forth, provided we do not lose sight of the fact that real social acts ontologically anchor them and exemplify them. Thus, there are already these multiple aspects of value to consider: (1) the perceived value quality as organically inscribed in affective life and dispositions with its significance or relative intensity (mutely enacted); (2) the name of a value quality (recognized-articulated); (3) the value quality as exemplified in social action, practice, or institution; (4) the "detached" ideal value quality as such, hardly distinguishable from the mere word, regarded as a norm or something that "ought to be"; and (5) the material conditions in which goods or moral values can be realized or not, which impose an "ought to do" when this "feasibility" exists. These are all up for discussion in various ways below.

Given the immense variability of human action and language, different acts can exemplify the same value, and different values can be used to describe the same act (relativist principle). But there is also far less variability than is usually assumed (Rolston's and ontogeny principles). Common embodiment and embeddedness in ecologies generates "vital values" that are relatively stable over the long term, in life span durations (survival principle). Before detailing these kinds of values, let me introduce some relations between these classes of value that will help us to understand just why our shared embodiment and embeddedness obligate communities to make vital values a priority for ecologically sustainable societies.

RELATIONS BETWEEN VALUE KINDS

Now, if we review some of the examples provided we can discern some relations between these two broad classes of goods and moral values. There seem to be two sorts of relation of dependence. In order to exhibit moral qualities, agents have to comport themselves among things and others with an eye to the goods values that those things and others bear. Environmentalists may strive to become better people by resisting dolphin slaughter in Japan, but only because they presuppose that the dolphins have a certain kind of value, are "subjects of a life" with "a good of their own."[11] This is a relation of *axiological* dependence between goods and moral values. The perceived values of things, goods, and situations, as disclosed in affective, embodied, cultural perception, orient human action in the world. Moreover, these values are not floating aloof in an isolated subjectivity disconnected from real ontological relationships, but are intimately interwoven with and dependent upon them. This exhibits their *ontological* dependence.

(1) *Axiological dependence* applies to one side of the relationship between the goods themselves and the human social and moral values. For example, there is theft—or honesty—only when things are experienced as valuable enough to steal in the first place; there is conservation or preservation of nature when some local region is deemed worth preserving. This is an "internal," axiological relation between goods values and moral values in general—whether or not the goods values in question exist. It is an "involvement" or "entailment" of goods values in moral values, an asymmetrical constitutional relationship of meaning or value.[12] We may also call this relation "intentional inclusion." This is a common feature of our everyday purposive behavior. In other words, we do something because we assume the end in view is "good," whether or not it actually is.

(2) At the same time, an *ontological dependence* between the bearers of goods and moral values also exists. The "internal" or axiological relation just described can be contrasted with an "external" relation between at least some goods and moral values. In the relation of ontological dependence, the

moral value does not "contain" goods as an element of value within itself, and the goods do not "recur" in the moral value. The conditioning values, for instance, allow the whole realm of moral goods and values to exist in the first place. This is the meaning of "constitutional relationality" already defined above. There simply would be no moral values at all without moral agencies embodied in vital subjects in a world of real objects, threats, promises, and social norms where survival and flourishing is a persistent goal.[13]

In other words, the dolphin activist "includes" the good or vital value of the dolphin in her compassion by axiologically presupposing it; but she can only exhibit compassion because she *is* a living, social, moral agent, on the one hand, and because the dolphin has (or is assumed to possess) a "good of its own" on the other. (Our attribution of "intrinsic value" to the dolphin is conditioned by the real "good for" the dolphin; our moral response depends implicitly on both.) To sum up the two relations here, we can say that *constitutional relationality is the ontological precondition for the intentional inclusion of goods values "in" moral values*. In other words, moral agents are living bodies directly dependent on environmental goods, and our disposition to "conserve" those goods, for example, "includes" them in the moral intention. Let's take another example. Cortland apples provide nutritional value only because, as living beings in real relations to them, we and other creatures can take advantage of their value. We "discover" rather than "project" this goods value. When one gives the apple to the neighbor's hungry child instead of eating it oneself, the goodness of the act is only partially dependent upon the real nutritional value of the apple, since if the apple turns out to be rotten, the moral worth of the act is not entirely annulled.[14] According to Rolston's principle, only certain bearers can carry certain values, although—given the indeterminacy of impulses and contingency of situations—a great range of values may be exhibited by any one carrier (relativist principle): apples are nutritious, but they may also have symbolic value as signs of fall, or projectile value in the giant slingshot competitions one sometimes finds at apple farms in New England. The internal axiological conditioning exists in the relation between the apple's "use" values and the moral value of the intention in giving it; and the constitutional relations between vital beings and sources of nutrition provide scaffolding for the moral relations of giving and receiving. More broadly speaking, when intentions toward widely approved environmental goods become stable and habitual dispositions, we can call them environmental (ecological) virtues. When a whole cluster of intentions aimed at patterns of value prioritization is reproduced in practices, institutions, communicative action and community, we call it an environmental (ecological) *ethos*. I will discuss these categories further in later sections. The following table summarizes the key points so far:

Table 4.1. Value Types and Their Relations

Type	Descriptor	Relation
Goods Values	Values intended, desired, striven for	Axiological: "included" in moral values Ontological: precondition for moral values
Moral Values	Values of the intention, desire, striving	Axiological: "includes" goods values Ontological: depends on goods values

To sum up, contrary to what is commonly believed given Modernist subjectivist metaphysics, values are not flimsy fabrications tied only to the whims of finicky valuers, but are the concrete contents of daily life. Even moral values like justice or generosity do not float in the air untethered to the Earth, but can be realized only in communities wherein unfairness avoided or goods freely given are already of communally recognized-articulated value or disvalue. Moreover, those values or disvalues can only be borne by certain really existing bearers, as not every object can have just any value indefinitely in the context of reproduction and development (survival principle).

A major outcome of this discussion of types of value and their relations is this: it shows that by tossing out the Procrustean bed of the instrumental-intrinsic value distinction, we can provide a more nuanced and powerful way of recognizing both the diversity and the commonality of social-ecological valuing. It allows us to see that these relations of axiological and ontological dependence provide another important criterion for a reasoned response to the problem of prioritization. It appears that the existence or realization of some values depends on the existence and realization of others, not only on the mere existence of potential bearers of value or of valuing agents. This would mean that some values must be recognized and actualized as conditions for the possibility of others. This is not a bizarre idea, even if the way it is expressed is somewhat abstruse. It is a point implicitly accepted even in neoliberal politics and ethics, where it is commonly assumed that equal opportunity is one of the highest policy goals and has to be secured before all else, because other values—like self-determination—depend on it. These relations of dependence thus provide a clue to prioritization, since if the higher (moral) values are desirable but depend on the lower, it is obvious that the realization of the conditioning values must first be achieved to render the conditioned ones possible at all. The distinction between goods and moral values arose from reflection on the structure of action, and the distinction will be important for approaching the question of prioritization. There is a subjective corollary to this distinction that can reinforce our sense of it even if it cannot provide a reliable criterion for responding appropriately to the problem of priorities: the intensity of our affective sense for values.

VALUE RELATIONS AND AFFECTIVE INTENSITY

There is no common denominator between goods and moral values; they are different in kind. They are not part of the same continuum or scale, even if they are necessarily related. The "external" ontological dependence relation expresses one kind of conditioning, while "intentional inclusion" includes the role that value feeling or affect plays in determining value rank or priority. The subjective correlate of the goods-moral values relation is the affective intensity of our response to value. Values are generally correlated with affects or feeling-intentions, and some theorists even claimed that there are very specific correspondences between specific values and particular emotions. While such specificity is unlikely (given the plastic and learned aspects of affect-responses), positive and negative affects certainly give us strong clues to socialized value ranking.

The status of conditioning, fundamental, or first-order value is disclosed in the negative experiences of disapproval, contempt, horror, disgust, and similar responses to their violation. The status of conditioned, complex, or second-order values is often revealed in experiences of assent, approval, acceptance, or enthusiasm for value. Experiences of disgust at the contamination of a "pure" sacred site or the serious disapproval felt at the violation of (negative) the human right to be free of harm in cases of torture reveals the *conditioning nature* of these fundamental values. This is also inversely reflected in the judgment that no one would be considered especially merit worthy for simply *not* defiling a sacred site or *not* torturing a prisoner—no moral merit accrues to the agent in such cases.[15] Thus, felt seriousness of offense is a good indicator of conditioning values, while degree of merit accruing to an agent serves as an indicator of conditioned (moral) value. These relations can be summed up in the idea that "the most grievous transgressions are those against the lowest [conditioning] values, but the greatest moral dessert attaches to the highest [conditioned] values."[16] It is certainly true that some goods can provoke stronger and weaker responses in us, and these responses are conditioned by all sorts of psychological and ideological factors. The second-order values in which we are primarily interested have their basis in a higher structural level, a normative level of approval and disapproval in light of how one "ought" to relate to such goods. There are thus both objective (dependence relations) and subjective (affective) criteria for recognizing what are fundamental or conditioned values. Unfortunately, the subjective side is somewhat unreliable because of our conative plasticity—food that may have been at one time appetizing can at another time be repulsive. Therefore, in order to pursue the question of rank ordering we have to rely primarily on the ontological-axiological relations highlighted above. But their affective correlates nevertheless provide a clue.

Another way to think about a rank order between conditioning and conditioned values is in terms of what I call the "sacrifice" test. With regard to your own value scale, what would you sacrifice in order to realize some higher value? We often hear about cases in which people sacrifice material possessions, luxuries, and some degree of comfort in order to preserve their health, and this indicates that health is more important to them than such goods. We also often hear the stereotype of the struggling artist or scientific genius who sacrifices their health for the sake of creating beautiful artworks or scientific theories, which expresses a specific cultural values-imaginary that ranks some forms of beauty or cultural knowledge above vital values (e.g., health). Finally, we also hear of martyrs who sacrifice their very lives for values that they think are beyond all other values. What the stereotypes obviously reveal is that we do conventionally and affectively rank values even when we are not aware of it. What does all of this have to do with environmental evaluation?

The central question for environmentalism becomes whether we can align our patterns of affective personal and communal rankings with real world relations of dependence between values and their bearers for the long term, in life span duration. Patterns of prioritizing in the long term can be called an *ethos*. An ethos, for an individual or community, is an enduring (habitual) set of motivations or cluster of value preferences that helps to define a pattern of practices within specific institutional, cultural, and ecological settings. In other words, it is patterns of prioritizing clusters of concrete value qualities that form the substance of individual and collective ethical life in developmental temporality. An ecological ethos would be one in which patterns of the prioritization of conditioning values became habitual and conventional. I call *vital values* a specific type of goods value that are both axiological and ontological conditions of moral values, like other goods values, but they are also a special type of condition without which there would be no valuing in the cosmos at all. If goods are goods "for" living things, then those living things have to exist for goods to exist at all. Living matter brings the whole dimension of constitutional evaluative relationality into being, and is its general condition. Vital values pertain to living things, processes, species, and even biotic communities or ecosystems. They can be regarded as a special class of goods that are necessary as (1) vital prerequisites for humankind's moral values as a whole (ontological dependence), and (2) as goods in relation to which specifically environmental as well as other moral values are realized (axiological dependence). Robustness, strength, vitality, health, and integrity are some vital values that permeate whole entities in a nonlocalizable and indivisible way. Vital values like biodiversity, integrity, and resilience of ecosystems are of special interest to the environmentalist because they express the vital processes of nonhuman nature upon which humans asymmetrically depend. Because vital values are conditions of all valuation, they are in a sense more fundamental than many other goods values, and some goods values are less fundamental than vital values. Useful material things, like building supplies, possessions, housing, energy sources, security fences, beds, and kitchen implements are physically extended and their value qualities are tied directly to their physical attributes as "use-values." These kinds of value are often localizable in things and divisible. All of these qualities can be experientially distinguished from another type of goods value, economic value, in whatever way this value is determined. As I will explain in the next chapter, economic or "exchange-value" is the result of a social calculus that goes on behind the backs of producers and consumers to which virtually no one involved has consented. Economic value has become an institutionalized social object that no longer has any direct tie to the physical or nonhuman natural world.

As I noted in chapter 3, while a number of writers provided a distinction between propertarian and priority questions, very few attempted to tackle the issue of prioritization or competing claims. Still, the need to deal with such conflicts was felt by a few authors. One obvious conflict in the literature took the form of the question "Can animal rights activists be environmentalists?"[17] There was an apparent conflict between emphasis on the value of individual animal lives and the holistic environmentalist emphasis on the value of the whole ecosystem, which considers some individuals expendable under some circumstances (e.g., wildlife management). Some authors, including Baird Callicott and Peter Wenz, went further to deal substantively with the question of the prioritization of competing claims. Callicott and Wenz attempted to deal with conflicting values by employing a concentric circle metaphor and prioritizing obligations to those "nearest" or with the "strongest" claims. Unfortunately, the rank ordering principles they invoke remain mostly untethered to embodied, social, evaluative experiences; they do not distinguish between goods and moral values; and because the relations between values and their bearers in the context of social reproduction and development remain hidden, their priority principles still remain open to the relativism objection.[18] What we wanted to know was just which values are higher or stronger than other values, and which ought to be prioritized in cases of conflict. Axiological value theory gives us a way of approaching priority questions without simply resorting to "our intuitions" or "conventionally accepted justifications," since in many cases what is commonly accepted is precisely anthropocentric and exploitative. Both models failed to provide satisfying accounts of prioritization. The relations between goods and moral values include relations of dependence. As I noted above, dependence relations define an order among values, and values themselves depend upon bearers and agents in specific ways. (The meaning of ontological dependence will be discussed further in chapter 6.) I elaborate on the principles of prioritization in the final section of this chapter.

PRINCIPLES OF PRIORITIZATION: STABILITY AND DYNAMISM

Recall the principle of dependence at the heart of environmentalism once again: humankind asymmetrically depends for its existence and flourishing on the nonhuman. The ethical correlate of this is that humans ought to preserve the conditions necessary for human *and nonhuman* survival and development. The axiological and ontological relations between goods and moral values explained above reflect both this ontological order of dependence and this normative claim.

To briefly review: the problem of priorities arises because the fundamental human situation in the world is one of finite human resources confronted by felt conflicts of values. We cannot give every moral claim its due, and moral life is life in the midst of conflicts of value.[19] Not all values are of the same sort, however. There are two basic classes of value defined by the structure of purposive action: goods and moral values. Humans are guided by enculturated patterns of valuation that sometimes respect and sometimes disregard the dependence relations between the two types of value. A social-naturalist perspective helps to mitigate the effects of dualism in human nature theory and the role of dualist logic in stereotyping motivation, and it also helpfully illuminates a developmental model for acquisition of valuational norms. All of these points lead to the question whether there is a scale or rank order of values that many different groups of people could agree gives reasonable priority to environmental values across many contexts, challenging the dominant ideological and anti-environmental modes of prioritization. I discuss these broader patterns further in the next chapter, and in this section add detail to the conception of prioritization based on the value dependencies discussed above.

The distinctions between goods and moral values and the two types of dependence relations give us clues toward the establishment of a scale of values relatively invariable at the conditioning level, as well as synthetic principles for guiding an ecological ethos. We said that goods values, or the specific qualities of things, are goods for living agents and persons in an objective sense of *for*. They are constitutionally relative to the capacities of those beings, and are no less real for being relational. Goods values (including vital values) are the necessary conditions for moral values.[20] Moral values arise in the intentions and actions of moral agents who are embodied and embedded in social-ecological interrelations. They thus depend ontologically on vital values, as well as on goods values of environmental and other kinds. The axiological dependence of moral values on goods and vital values means that certain vital values must be present for moral agency to exist at all; in other words, certain vital values and goods must be actual in order to realize certain kinds of environmental virtues in the agent. A general principle evidently follows from these relations of dependence: *One ought not consistently value (in developmental duration) the necessary condition for producing, reproducing, and developing human and nonhuman life less than one values the developed life itself.* In other words, since many goods values depend on vital values as their conditions (and moral values depend on both), *persistently* backgrounding or denying the value of the conditions becomes a moral error. (This is one aspect of the feature of backgrounding or denial of dependency that Plumwood defined in her analysis of dualist logic.) For instance, an anthropocentric mode of valuing that prefers profit to ecological integrity is thus in moral and practical contradiction with itself, provided we take the life span (production and reproduction) of human and nonhuman communities into account. Widespread recognition of this current pattern of misprioritization could motivate an ethos of sustainability. It expresses a priority of the conditioning values (and valuers). Such reasoning necessarily presupposes the life and existence of real nonhuman valuers (subjects of a life, living things with a good of their own), and this value is both intentionally included in prudential reasoning and these nonhumans are its ontological condition in life span duration. Therefore, we can say that social systems that violate vital values necessary for the production, reproduction, and development of human and nonhuman life are in practical contradiction with themselves, and must be transformed. That is, if what every social, economic, or political system aims at is the reproduction and development of human and nonhuman life, and yet its current organization systematically destroys, exploits, or otherwise harms vital values (and exploits humans), then it is accountable when it fails to do so. It is in "performative contradiction" with itself.[21] More importantly, any ethical system that is embodied in a political economic system that does not allow human life to survive and develop, that produces human and nonhuman victims, is self-contradictory at a collective level. Such systems have to be criticized and transformed for the sake of liberating the victims. Plumwood's liberation model of anthropocentrism allows environmentalists to join the struggle with other liberation movements who confront a common enemy in the system of white supremacist, patriarchal, racist, speciesist, classist, capitalist society. I will address this systematic level in the next chapter.

Explained in more detail, the root of this contradiction lies in the relation of ontological dependence defined above. Axiologically, the relation between goods and moral values is not scaled or direct. This means that the value of a good does not correlate with the moral value that intentionally includes it. For example, the moral merit accruing to the sacrifice the homeless person makes who gives their last crust of pizza to another, despite their own hunger, is as great as if they gave away a treasured heirloom. The value of the sacrifice is not altered by the fact that mere pizza rather than an heirloom was given. The intentional inclusion is preserved in the relation, but otherwise the conditioned (moral) value is axiologically not directly affected by the goods value on which it depends.[22] Ontologically however—and here is the crucial point for environmentalist prioritization—there is a correlation between goods and morals in terms of existing value conditions. This is because conditioning pertains primarily to the *actualization* of one value and *actualization* of another. If one value has to be concretely actualized (in duration) before the other can be, we have an ontological dependence relation, not simply one of axiological inclusion. This is the key to environmentalist prioritization. Thus, moral values such as Taylor's "respect for life" correspond to the most fundamental conditioning environmental values, the life and existence of nonhumans.

Conditioning goods values correspond to "stronger" moral values. Justice is a stronger or conditioning moral value because it implicates the more elementary goods in itself, like property, access, participation, and material goods. "Good taste" is a more conditioned value and implicates "spiritual goods" like beauty, which is why offenses against taste may be disapproved but are not punishable. Worth of the moral value and of the goods need not correspond, and often do not, for moral merit; but value of the good does correspond to conditioning moral values, since violation of the conditioning is more strongly felt (affective intensity).[23] Respect, for example, is properly limited to other agents, human or nonhuman, or to spiritual goods directly bound to agents, like cultural knowledge or traditions. Taylor's "respect for life" is thus destroyed when life is destroyed; the possibility of a moral value that ought to be part of the *ethos* of the environmentalist is eliminated. In such a case, the strength of the goods values does matter. The right kinds of goods (e.g., living things) must evoke the right kind of complementary response (e.g., respect) within the larger ethos. It goes without saying that where living things are not even recognized as elementary vital values (and valuers), but simply as means to achieve some other goods values, there is not even the possibility of an environmentalist ethos. While this valuational error might be described as a matter of subjectively treating something as an instrumental value when it really has intrinsic value, it is better understood as matter of being blind to the real structural relations of dependence between the values (life and respect), and so between the carriers of value themselves (living things and persons). Stronger (in the sense of conditioning) values have a more invariable claim. These vital values are similar for most people because they make up humanity's constitutive ecological relations, and are neither strictly biological (needs) nor strictly cultural (interests). Vital goods and stronger moral values are correlated; higher values are not correlated with specific goods values. Stronger vital values as constituents of growth, reproduction, and survival for flourishing human and nonhuman communities must be respected, given the principles of dependence. Valuing such ecological goods (and our own vitality in concert with them) gives an aspect of stability to an ecological ethos.

However, too much emphasis on conserving the conditioning values narrows our sense of what is genuinely moral about environmentalism. As far as we can tell, we are the only being capable of the moral value of planetary care. If our task is to realize values which without us would not exist (I can't defend this claim here but will take it as granted), then this is certainly a value unique to humankind. It is what drives environmentalism forward, where battles over conservation and preservation are seen as necessary but not sufficient for fully developed human and nonhuman life. "Dynamic" values are those that are inspiring, beautiful, admirable, merit worthy in human action. Vital values are irreplaceably necessary, but a life with only the simple conditions satisfied, without admiration, without sacrifice, without kindness and compassion, without solidarity, self-development and self-realization—that cannot be estimated a developed human life. Such second-order values should also have priority where developing human life is our goal. There is thus also a "priority of the conditioned." The second principle of prioritization is that *one ought to foster, cultivate, and develop the dynamic values by aiming at the right kinds of goods.*

These two apparently opposed directions can be synthesized in a principle of stability and dynamism. These observations suggest that the aim of moral action in the midst of conflicts should be to integrate conservative and dynamic moral impulses at once.[24] The third principle of prioritization then encourages a double preference for the right kinds of goods and the right kinds of moral values of the agent. On the one hand, there is our appreciation for merit or praiseworthiness. On the other, there is our degree of felt violation or offense. For example, murder and enslavement are felt to be deeply criminal because they offend against the most fundamental conditioning values of life and autonomy. People are not imprisoned for not being courageous or not promoting sustainability, and this indicates that these virtues belong to a more content-rich but "weaker" stratum of moral value. Being generous is more meritorious or deserving of praise than simply refraining from stealing. One is a "better person" for being forgiving, just, and wise, but if one lacks these traits it is not considered a crime. The conditioning values are more strongly responded to, while the second-order values have richer (and more variable) content but are weaker.

The third general principle can thus be expressed as a synthesis of the first two: to be good one has to nourish and not offend against the foundational values and at the same time cultivate second-order values (partly by means of conserving vital conditions for humans and nonhumans).[25] On this view, the ideal human ethos is a creative synthesis of these two ethical preferential tendencies. If "genuine morality is built from below up," and "only upon the actualization of the lower does the actualization of the higher rest solidly," then the moral ideal must in the end aspire to respond to the demands of both the conditioning and the conditioned values. We know that without vital values, flourishing human and nonhuman life is not possible. Yet, if the human task in the world is to realize values which without us would not exist, then moral values are precisely those we dynamically aspire to realize beyond merely ensuring the existence of vital values. Environmental ethics on this account would be one in which the moral qualities of the person stand in particular sorts of relation to these vital values, such as relations of recognition and support (enabling them), and resistance to practices and institutions that degrade the conditions of vital values. If it turns out that business-as-usual individual and community patterns of behavior degrade vital human and environmental values, then environmentalists have to be opposed to them.

If this is a reasonable description of how things stand, then we have to take seriously the immense inequalities of wealth and power that exist between human communities on the planet, and the way these inequalities also impact *nonhuman* lives and communities, since we cannot be ourselves without them. For ecological and social sustainability we must enforce the trend to not offend against the vital conditioning values of biodiversity, ecosystem health, integrity, life, and its provisioning. But this is a merely negative or conservative prerequisite; it is not the supreme goal of environmentalism. The point is to become the fully developed personalities in the collective human and nonhuman communities we are capable of becoming. This is why sustainable development in its function as maldevelopment, expropriation, impoverishment, and monoculture export is so reprehensible.[26] It takes away not just the conditioning, situational, and goods values of human life, by doing so it makes it impossible to foster comprehensively the more meritorious values of personality, beauty, knowledge, and genuine community in a reproducible way into the indefinite future. We currently strive to achieve global environmental justice, but justice (like equality) is one of the lowest of the moral values. Love of one's neighbor (including nonhuman ones), courage, honesty, trustworthiness, fidelity, generosity, personality, or self-realization all exceed justice in moral merit, and give flourishing human life its depth and dynamic richness. The obvious requirement is to recognize that moral values are *dependent* upon life, both within and without the subject or person. If life is materially-axiologically essential to any moral value, then we ought to preserve it. This means often prioritizing it over many kinds of human interests in environmental decision-making.

In terms of the classic categories of anthropocentric, biocentric, and ecocentric environmental ethics, this theory qualifies as none of the above. This is because the concept of centrism on which they are based assumes the idea that there must be a single supreme value as organizing principle of prioritization (usually defined in a propertarian way—subject of a life, intrinsic value, wholeness, rights, etc.). If we reject propertarianism, we must also reject centrisms of all kinds. Even multicentrism, although an apparent gain, is still defined in a negatively propertarian way.[27] We have to do away with properties altogether as organizing factors, and instead talk about priorities. There are no singular dominant values here. My view may resemble Norton's weak anthropocentrism because it does not discount human prudential interests in environmental decision-making; it also embraces Rolston's ecocentrism because it acknowledges the self-valuing and value-producing capacities of living and nonliving systems; and it incorporates feminist, pragmatist, and virtue ethical approaches that place emphasis on an ethos of openness, courage, justice, kinship, and environmental engagement. These approaches take on a new aspect as soon as we take the problem of prioritization to be central to our ethical lives. Inserting all of this into the context of political ecology will be the task of the next chapter.

All of this allows us to respond with some guidance to the problem of prioritization. The first principle states that one ought not *consistently value (in developmental duration) the necessary condition for producing, reproducing, and developing human and nonhuman life less than you value the developed life itself*, since some values are conditioned by or dependent upon others. It acknowledges the priority of the conditioning values. The second principle states that *one ought to foster, cultivate, and develop the dynamic values by aiming at the right kinds of goods*. The third principle integrates and comprehends these by urging us *to nourish and not offend against the foundational values and at the same time cultivate second-order values (partly by means of conserving vital conditions for humans and nonhumans)*. For an ecological ethic, *goodness* means nourishing and not offending against the foundational values while at the same time cultivating the dynamic values in a creative synthesis of these two ethical preferential tendencies. Imagine a world where individuals in communities participate in ongoing processes of negotiating the conflicting values that arise in the course of leading their lives, assessing what are the stronger community needs and enabling higher individual self-realization. None of this happens without formal or informal institutions. We now have to consider value prioritization in the context of dominant institutional frameworks, where consistent patterns of acknowledging dependence are juxtaposed with patterns of ignoring it in capitalist cultures. The task of prioritizing intersects directly with political ecology, since we can see that only community-level communicative action is adequate to prioritize values or goals that are in the interests of the human and nonhuman community.

CHAPTER FIVE

Political Ecology and Value Theory

ECOPOLITICS AND ETHOS

The term *political ecology* refers to at least two distinct sorts of writing. In the empirical social sciences broadly construed, such as anthropology, development studies, environmental sociology, and geography, it usually means careful political, economic, and environmental analysis of the social forms of access to and control over natural resources and their implications for environmental health and sustainable livelihoods.[1] In philosophy, the term *political ecology* was proposed by philosopher John P. Clark as a more inclusive substitute for the term *social ecology* made familiar by Murray Bookchin.[2] As such it would accommodate the perspectives on human-nature relations offered by theoretical environmental justice work, anarchist social ecology, ecosocialism or Marxian ecology, and many forms of ecofeminism. Both approaches to political ecology share the central insight that human exploitation of nature is directly linked to human social domination. If the ecological crisis is in large part a result of human "domination" of nature, and "the idea of dominating nature has its primary source in the domination of human by human," as Bookchin put it, then the environmental crisis is also a social crisis.[3] In other words, environmental injustices are inevitably bound up with social injustices. Political ecologists also usually share the recognition that *humanity* and *nature* are categories that are far too baggy and indeterminate to engage in work that requires explaining how some human groups exploit others, or exploit resources in a way that creates differential harms and benefits for different communities. These two tenets—the tight conceptual and empirical linkage between social hierarchy and environmental exploitation, and the necessity of dehomogenizing the classes *human* and *nature*—will serve to minimally characterize political ecology here. These readily distinguish it from a tradition of environmental ethics that acknowledges neither of these points. It should be obvious that Plumwood's critique of dualisms falls under the umbrella of political ecology, and gives it a substantial philosophical framework.

So far, I have argued that ethics is a matter of social and affective cognition increasingly refined in the ontogenetic development of unfinished animals requiring orientation amidst an abundance of values (or valuable things, relations, and situations) in their natural-social environments. While the agential perspective is crucial for discussing motivation, motivations also have to be conceived in terms of their recurrent circuiting through past and current social practices and their consequences. Moreover, these have to be understood to be part of a social institutional complex that may be seen not only as disciplining practices and molding plastic motivations, but as infused with imaginary or ideological components that allow hierarchical or other power relations to maintain their sway and attain a high degree of stability. It is the power relations studied by empirical political ecologists that are said to provide access to and control over natural resources, and these cannot be separated from their social institutional, ideological, and imaginary aspects. Dominant ideologies play themselves out in social institutions and practices, and the multiplicity of values has to be regarded as embedded in an already pervasive social institutional matrix.

The concept of *ethos* (briefly mentioned in the previous chapter) brings together motivations and practices, individual and social, intention and institution. The dimension of ethos, extended through developmental temporality, should be considered the nexus of individual and social determinations. Through the deliberate combination of agent-centered and political ecological perspectives we may gain a more comprehensive conception of an environmentalist ethos. An ethos, for an individual or community, is a persistent, habitual set of motivations, dispositions, or cluster of valuing tendencies that helps to define a pattern of practices within institutional, cultural, and ecological settings. Ethos emphasizes the centrality of duration, temporality, or lived time that shapes the relations between conditioned and conditioning values and their bearers. It is the dimension in which growth, development, and survival for self and others are assured or undermined through patterned prioritization of goods and moral values in social institutional contexts.

Political ecologists have generally neglected this dimension and focused most attention on other aspects of social determination—such as its extensive and indispensable ideology critique. Therefore, a political ecological analysis that emphasizes this dimension of ethical experience is warranted.

Put simply, values talk must ultimately be placed within the larger social context of institutionalized capitalist political economy. Ecosocialists and their allies have been particularly good at analyzing the negative environmental impacts of capitalism as a political economic and cultural system, but have not done as well at conjoining ethical and political analysis. After introducing ecosocialist Joel Kovel's attempt to do so, I touch on "ecosystem services discourse"—the popular framework for economic valuation and potential commodification of nonhuman nature—to show how propertarian institutions and ideology continue to shape environmental prioritization and policy proposals in negative ways, and to clarify the place of economic value in the value taxonomy of this book. This sideways glance at ecosystem services discourse should not be regarded as an exhaustive critique and evaluation of its vast literature, nor as an attempt to dismiss it entirely. My goal is limited to making the specific claim that ecosystem services discourse perpetuates exploitative relations between many human communities and nonhuman nature because its pattern of systematically prioritizing certain goods over vital and moral values is shaped by propertarian institutions and practices. This last chapter on value theory in part 2 of the book aims to integrate the range of axiological issues previously covered with recent critiques of consumer capitalist culture's environmental destructiveness. In value theoretical terms, I conclude that the capitalist ethos violates the principles of prioritization, the conservative "priority of the lower" and the dynamic synthesis of preservative and creative tendencies. Resistance to its conception of progress and growth should entail a critical ethical consciousness that emphasizes the safeguarding of vital values and the dynamic development of self and collective human and nonhuman community through enabling environmental practices and more deliberate consensus-based prioritization.

FORMS OF SOCIAL DETERMINATION

In a recent book, John P. Clark introduced an effective heuristic for political ecological analysis within which this examination of contrasting *ethe* may conveniently be situated.[4] Clark has been a part of environmental philosophy and politics since at least the mid-1980s, and is probably best known as co-editor of a popular environmental philosophy textbook, the section on political ecology and long introduction to which is one of its best features.[5] His recently published monograph *The Impossible Community: Realizing Communitarian Anarchism* (2013) contains a remarkably wide-ranging array of political ecological insights, and in it he begins to develop a "dialectical social ecology," designed to account for the ways that the four spheres of social reality—the social institutional structure, social ideology, social imaginary, and social ethos—dialectically interact to produce varieties of social determination (a general concept encompassing the various ways in which human social behavior is conditioned and shaped). I will briefly sketch Clark's conception of social determination, especially highlighting the role of what he calls the social ethos, and will examine its significance for this analysis.

The two broad themes of Clark's book include an "inquiry into the ways in which social transformation is possible, and the analysis of the goal of such transformation, which is the free community."[6] Social transformation for the sake of averting global ecological catastrophe requires an innovative theory of social determination "that encompasses an analysis of the spheres of the social institutional structure, social ideology, the social imaginary, and the social ethos" that I believe is more comprehensive and fruitful than other political ecological approaches to social determination. The first two spheres are given most attention in leftist social analyses.

(1) The *institutional structure* includes the structure of capital, the state apparatus, the technological system, the formal structure of social practices, determining rules of the system, and "the actual structuration of material constituents and resources in accord with such principles." This includes "institutional systems of domination and oppression based on sex, race, sexual orientation, culture, ethnicity, etc."[7]

(2) These material structures are directly reinforced by the *social ideology*, defined in standard critical theory terms as "a system of ideas that purports to be an objective depiction of reality but in fact constitutes a systematic distortion of reality on behalf of some particularistic interest or some system of differential power."[8] The dualistic logic of domination identified by Plumwood is one form of such pervasive social ideology.

(3) In addition, one has to consider the society's "collective fantasy life," or *social imaginary*. "It includes socially conditioned self-images, commodity images, and images of the other. It includes the prevailing myths and

paradigmatic narratives.... It is related to certain preeminent institutions of the imaginary, such as advertising, marketing, mass media, the arts, and mass culture in general."[9] One of the prevailing myths in current society is that technological and economic growth is potentially infinite, and its technocratic optimism often fuels science-driven environmentalist responses to crisis.

(4) Finally, according to Clark, the "social ethos, habitus, or structure and content of practice in everyday life" is the "most neglected" sphere of analysis among leftists, but also "the *most crucial area for the establishment or transformation of patterns of behavior and forms of consciousness*."[10] As "part of the collective subjective dimension of the dominant system, ... ethos encompasses the prevailing cultural climate of a community or society.... Ethos is the sphere of social psychological reality."[11] With reference to Hegel, Clark also calls this sphere that of "ethical substantiality," which is embodied in the "histories, values, and practices of evolving communities and cultures."[12] On his dialectical approach, of course, each of these spheres is only analytically distinct from the others, and their constant interactions and complex relations have to be mapped in any account aspiring to completeness. The point of the analysis of the four spheres is not merely to better understand social determination, but "to understand what must be done if patterns of thought and action are to emerge that truly challenge and begin to overturn the system of social domination."[13] In this book, I have taken Clark's claim that ethos has been most neglected to heart and have aimed to remedy this by developing a value theory that gives more specific content to this "most crucial area for the establishment or transformation of patterns of behavior and forms of consciousness."

As noted above, ethos conjoins the phenomena of cultural tradition, embodied practice, and patterns of valuation. In chapter 2, I noted that Tomasello emphasizes the ontogenetic timescale where every individual in the culture develops and is initiated into the world of social-cognitive intentionality and perspectival attention. I said that if environmentalists want to understand and transform current behaviors that destroy or preserve ecosystems, and if the cultural norms, conventions, and habits (social ethos) of a community are the learning environment of human children from birth, then to pass over the ontogenetic timescale—the proximate developmental context of human life—is a big mistake. Since culture is the "species-typical and species-unique 'ontogenetic niche' for human development," it is where "the particular habitus into which a child is born determines the kinds of social interactions she will have, the kinds of physical objects she will have available, the kinds of learning experiences and opportunities she will encounter, and the kinds of inferences she will draw about the way of life of those around her."[14] We can conclude that if the proximate cause of current antiecological behavior is such cultural patterns of choice in institutional contexts, and the proximate context of patterns of choice is the ontogenetic natural-social environment, then transforming these patterns requires understanding the embodied, social-natural, affective-cognitive patterns of valuation expressed in the social ethos.

Clark accordingly claims that social structures "function through the imposition of social values." He continues:

> To the extent that members of a society participate in these structures, they internalize the dominant values embedded in the structures (whether they affirm these values explicitly or not, as they usually do not). What is at least as significant is that they enact those values through their social practice. Thus, they reproduce the dominant values both through their acts of mind and through their social acts. By means of their very ubiquity (pervading social institutions, the social ideology, the social imaginary, and the social ethos) the dominant values take on a powerful moral force, achieve a high degree of self-evidence, and thus constitute *ethical reality* for the members of the society.... Social value is systemic in nature and there exists a dominant system of values.[15]

Consistent with the value theory developed above, Clark suggests that an ethos is thus constituted by certain patterns of value attentiveness and appreciation, evaluations, and judgments that are expressed in practices. An ethos has both first and third personal aspects, since observers can ask what values and disvalues an agent appears to be promoting and avoiding in behavior (facilitating third-party norm enforcement), just as the agent can ask herself in moments of second-order reflection whether the ethos she lives is a good one or not. We might ask ourselves "Were these wants implanted in me by an ideology, or are they really mine?" "Am I a good person for pursuing these ends rather than others?" "Am I complicit in the system of human and nonhuman domination?" Given our overall behavioral conservatism, powerful inclinations to act as our friends act, and to maintain smooth social relations, we often end up just reaffirming the old patterns. The problem becomes one of reproducibly interrupting those patterns to create the kind of long-term changes invoked by writers like Clark, Plumwood, Gare, and others.

In an important example, Clark discusses the pernicious function of commodity fetishism in patterns of valuation. Derived from Marx's analysis of capital, commodity fetishism is the idea that commodities intrinsically possess economic value, although as Clark explains, it also has an extended sense.

> A social agent may assert that the highest good for him or her is, for example, religious faith, moral integrity, family values, the sanctity of life, individual freedom, or some other values. Yet, at the same time, this agent may devote the vast majority of his or her waking life to producing and consuming commodities, may invest more psychical energy into the commodity than in any supposedly higher good, and may regularly decide conflicts between other values and commodity values in favor of the commodity. This is, in fact, what occurs regularly in liberal capitalist societies.[16]

The significance of this insight cannot be overestimated. The passage implies that we persistently experience conflicts between plural values, and it gives the problem of prioritizing values a central role in human experience. The passage recognizes the function of dominant ideologies, imaginaries, and institutions in shaping our patterns of prioritization. At this stage, the central elements are in place for a meaningful discussion of ethical substantiality—including an account of social determination, a naturalistic anthropology, and value pluralism—but what must be provided is a taxonomy of values themselves and a discussion of their relations that would supply some guidelines for the resolution of conflicts in favor of ecological rather than capitalistic values. The taxonomy presented in the last chapter fills precisely this role.

Clark claims that classical anarchism has always emphasized "self-realization and self-determination," which "makes such value pluralism an inherent part of freedom."[17] On my view, self-realization can be taken to be a significant moral value that requires certain goods and vital values as its conditions. While self-realization is indeed a pivotal moral value, promoting it is not enough to ensure that the right patterns of prioritization for reproduction, growth, and development will prevail in the long run. This means that we must explicitly embrace the conception of value that builds on the natural-cultural embodiment of vital, moral agents. For instance, with reference to that framework we could argue that self-realization necessarily entails that some goods values exist as condition of its process of achievement (assuming that this process is never completed). The axiological relation of dependence between moral and goods values does not say exactly what these goods are, nor what they should be for any individual in community. However, the values of goods also depend ontologically on the material constitution of their carriers. Thus, if we say that human and ecosystem health (and in turn biodiversity, clean water and air, etc.) are conditions of self-realization, we assume the concrete *agency* of biotic and abiotic components of ecosystems and the vitality of the moral agent as ontological conditions for vital values. This is an invariable relation of dependence where, in a conflict of values that pits one value or cluster of values against another (such as profit against personal and environmental health), to *persistently* opt for profit over health in life span duration has to be regarded as a moral violation and offense against the stronger conditioning values of life and health. A culture that ideologically steers preference and choice in such conflicts in the direction of commodities is thus morally flawed because it systematically undermines the real conditions for the achievement of flourishing human (and often nonhuman) lives.[18] To reiterate the first principle of prioritization: One ought not *consistently value (in developmental duration) the necessary condition for producing, reproducing, and developing human and nonhuman life less than one values the developed life itself.* There is a priority of the conditioning values. Another more familiar example of this relation: if you value your gas-powered automobile, you cannot consistently (in life span duration) devalue gasoline or ultimately the vehicle becomes useless. Since goods values depend on vital values as their conditions—and moral values depend on both—backgrounding or ignoring the value of the conditions becomes a moral problem. There are invariable structures or relations of dependence in the developmental (social-ecological) context that support individual growth (self-realization) and reproduction of human and nonhuman life in communities. Those structures are necessary conditions under which beings capable of leading a life on their own initiative are reproduced. Relations of dependence among values and their bearers are reflected in these structures, and while a high degree of variability is characteristic of some types of value, others are shared as a consequence of common embodiment and embeddedness in ecologies.

A nonsubjectivist counterideology has to insist on the relative stability of real-world ecological relations and their indispensable value for human and nonhuman life. Such relations and processes do not have value because they are desired, they are desirable because they are constitutionally relational values for living beings, such as ourselves, whether we in fact value them or not. As Clark's fetishism example shows, if the question is not only what is valued, but which values are systematically prioritized over others, and that some patterns of prioritization are ideologically shaped, it implies that there is a "right" pattern of prioritization and a "wrong" (ideological) one if building sustainable communities is the goal. This means that there must be some fundamental shared values in an invariable order scaffolding even participatory deliberative processes (in which we might engage in order to achieve social change) themselves. The patterns of prioritization, or norms characterizing an ethos, should be "creative products of processes of participatory social determination," even if they often are not.[19] Every norm formulation is based on the prior existence of values of various kinds tethered to real ontological and axiological relationships of dependence. This means that we can have relatively stable values that are shared without the immediate danger of pernicious universalization or imposition since patterns of prioritization should ideally pass through critical processes of communal participatory social determination. To counter the patterns of commodity fetishism, then, we have to articulate the ways that most of the higher values that persons consider aspects of self-realized, flourishing human lives are axiologically dependent upon conditioning goods values, especially those that we call vital values of the subject herself as well as the living processes surrounding the subject in her ecological embeddedness. The value scheme presented in the last chapter made this order of dependence evident. The simple observation that both humans and other-than-humans depend upon clean air, water, and nontoxic surroundings in order to reproduce and develop also reflects this order. The problem with the dominant system then is that this ecological and axiological order of dependence has been systematically subverted and is continually subjectively denied within an ecocidal culture. As Plumwood noted, in it "dependency on nature is denied, systematically, so that nature's order, resistance and survival requirements are not perceived as imposing a limit on human goals or enterprises."[20] Political ecology has revealed the way that ideological and institutional economic valuation has shaped human treatment of nonhuman nature, in both crude and subtle ways. To see more clearly exactly how this happens in terms of patterns of value prioritization, we should consider the place of economic value in the value taxonomy developed here.

ECONOMIC VALUE AND ECOSYSTEM SERVICES

Ecosocialism is one school of thought in political ecology and activism that sheds light on economic value within the capitalist system relative to environmentalism.[21] According to the late Joel Kovel, capitalism is the root cause of ecological degradation, and unless global humanity entirely rejects capitalism as a way of life, it will bring about "the end of the world." He argued that capitalism's ruthless commodification of every aspect of life has deformed human relations to nature. Its exploitative social organization of labor prevents the free association of laborers and their creative expression, just as it denies agency and autonomy to nature. Therefore, an ecosocialist society must be designed to establish new ways of organizing labor (some of which already exist) and a new attitude toward production in relation to nonhuman nature. Kovel calls *ecosocialist* "that society in which production is carried out by freely associated labor and with consciously ecocentric means and ends."[22] Admittedly, traditional socialism has not been especially environmentally conscious in its history, and must become ecosocialism or it risks perpetuating human separation from and degradation of nature characteristic of the socialist heritage.[23] According to him, "production" must become "ecocentric production," which includes valuing process as much as product, freely chosen labor (as against capitalism's coercion), mutual recognition, acceptance of "limits to growth" both in terms of ecosystemic energy and of human needs (sufficiency), appropriate technology directed to generating healthy ecosystems rather than profit, and "receptivity to nature" in its fullness. Ecosocialism must be an international movement to confront global capital, even if its initiatives are necessarily local responses to regional problems. Kovel thought the problem of global climate disruption might mobilize humanity and generate a cultural tipping point toward large-scale adoption of ecocentric production and establishment of ecosocialist communities. As he saw it, the ecosocialist movement forms a fertile matrix of freely evolving, participatory, solidary initiatives founded in bioregional communities of resistance and production.[24]

For the purposes of my project, it is interesting to note that Kovel also claimed that "ecological politics can be translated into a framework of values."[25] In a telling passage, he wrote that

> ecosocialism entails different and more complex judgments of *value* than first-epoch socialism. It demands of us that we take into account a kind of valuation distinct from those values, attached to use and exchange, that enter into economic calculation. Once we open ourselves to the ecosphere, a realm of *intrinsic value* opens as well, a value inhering in ecosystemic being.... This perspective becomes necessary in the overcoming of the ecological crisis, and therefore the climate crisis as well. It is the deepest level of the resistance to capital, and the foundation of all others.[26]

As I have argued, articulating one's values, and those of the community one inhabits or strives to inhabit, is the first step toward more deliberately actualizing them. Setting aside the reference to intrinsic value here, in Kovel's comprehensive perspective this entails becoming actively receptive to nature's agency, fecundity, formativity, respecting life and persons, and embracing sufficiency and sustainability as opposed to consumption, wastefulness, and incessant growth. Kovel's work might be viewed as an example of the type of creative synthesis of higher conditioned and stronger conditioning values discussed in the previous chapter, as well as an example of a critique of the kinds of debilitating social institutions, social ideology, and social imaginary discussed by Clark. The values discourse is not meant to replace but to complement the analysis of power and of structures of constraint and enablement configured by capitalism as discussed elsewhere in his work and in political ecology more broadly. It does not treat the environmentalist as a moral agent abstracted from concrete social relations, but effectively supplements social scientific accounts of that agency with a social psychological dimension that is indispensable for better understanding our own and others' motivations, and just as importantly, for envisioning and implementing change. Critiques of the current system are obviously indispensable, but they remain purely negative without a positive account of how social agents can engage in transformative change. For undertaking both, it will be helpful to briefly consider his definitions of key terms such as *exchange-value* and the way he understands the implicit prioritizations of economic value entailed in capitalism.

Kovel's most extensive discussion of the different types of value and ways of valuing nature took place in the core chapters of his book *The Enemy of Nature*. Kovel claimed that "intentions are deployments of values."[27] For anti-capitalist movements, this means that intentional communities that strive to extricate themselves from the "capitalist force field" would prioritize values other than those inculcated by consumerist culture. Insofar as Kovel speaks here of intentions at all, he is sensitive to the distinction between goods (values intended) and moral values (value of the intention) explained above. When his ecosocialist actors consciously intend to reject commodity fetishism, immediate gratification, and convenience, for example, and embrace instead ecological integrity, wholeness, and sufficiency, he tacitly acknowledges that the moral worth of the intention to realize these values is greater than the intention to "reduce the world to cash."

Kovel begins with Marx's distinction between use-value and exchange-value.[28] He says of both use-value and exchange-value that they are "qualitative and profoundly political." For Marx, objects may have a use-value or be useful for some purpose without ever becoming commodities (which possess both use- and exchange-values). Their qualitative usefulness is conditioned in part by their physical constitution, as well as by our own constitutional needs and the labor process. Thus, use-values are "relational properties" but never exhaust what an object might have in store for us.[29] This is the "for" of "constitutional relationality." A whole host of qualities can be captured by the category of use-value. The kind term *use-value* is analogous to the term *instrumental value* used by environmental ethicists: they both represent the genus of many species qualities that fall under them, and do not themselves express one specific quality or value. Use-values might include the water-deflecting quality of an umbrella or the rigidity of a cardboard box. Specific qualities such as these—though not all use-values—are also the results of qualitatively distinctive labor processes (e.g., weaving or corrugating).

Exchange-value is Marx's term for the average cost of socially necessary labor time to produce a given commodity. It represents a quantity of value in contrast to the specific useful qualities of things. Exchange-value is the quantitative side of the commodity and is derived from the network of social relationships of production and consumption rather than from the constitution of things themselves. The role of price is to represent a quantity of exchange-value.[30] In my terms, exchange-value should be regarded as a kind of goods value—one whose specific quality is to be a measure of quantity. Once this type of value is abstracted from the sensuous and concrete world of use-values and ecosystem integrity, it can expand under a political economic system oriented to wealth accumulation and create a new order of economic value.[31] These two whole orders of value are incommensurable, and this Marxian insight is the basis for the further insight that the capitalist order of exchange-values (the market economy) and the order of use-values (ecosystem and human ecology) are fundamentally at odds. As Kovel noted, capitalism sets up "an alternative, monetized world, with no fixed connection to the original world."[32] Once this happens, disintegration of communities and ecologies prevails. This separation of exchange-value from use, producers from means of production, human beings from nature, "is the fundamental gesture of capital."[33]

Now, what good are these economic categories for developing an "ecocentric ethic"? It is a question of priorities. Under capitalism, an all-consuming prioritization of exchange-value reigns, whereas in a liberated ecological society, the emphasis would be placed on use-values. Kovel said that "under normal capitalist conditions, exchange-value prevails and use-values are subordinated and degraded." Again, "capitalism comprises the society which sees to it that xv>uv [exchange-value takes precedence over use-value] so that people internalize the signals of the market and obey them as gospel."[34] Kovel recognized that individual humans as well as whole societies can be disfigured by the obsession with one value at the expense of others. In Clark's terms, the widespread commodity fetishist social ideology and social imaginary encourage most people to see exchange-value rather than intrinsic or use-value in things and environments. However, "in a liberated and ecologically sane world, use-values would take on a character independent of exchange-value . . . to serve the needs of human nature and nature."[35] In order to motivate this change—I would add—it is necessary to specify qualities or values-to-be-realized in contrast to those promoted by the capitalist system. Kovel strives to articulate these alternative values with an appeal to a wholly un-Marxian category of intrinsic value. It should be clear from all of this that choosing to value use-value or intrinsic value over exchange-value is something individual persons do in the context of the natural and social worlds opened to their perception and experience, already structured by social imaginaries, ideologies, and institutions.

What the general political ecological reorientation offered by Kovel and Clark reveals is that our evaluative processes are embedded in the context of commodity fetishism and growth ideology, and these are supported by institutions bound to reproduce these same patterns of valuation. In contemporary industrial and consumer capitalist culture the axiological and ontological dependencies discussed in the last chapter are masked by forms of social determination that make such connections invisible. Propertarianism is a core part of the political economic institutions, ideologies, and imaginaries that frame ethical relations to nonhuman nature in many societies. The questions "What properties does it have?" and "Whose property is it?" are both shaped by the dualist logic of domination identified by Plumwood, as well as by traditional Modernist substance metaphysics that takes discrete individuals to be ontologically primary and relations secondary. On the dominant anthropocentric model, only persons (defined as possessing rationality, autonomy, etc.) are both moral patients and agents, and the person-property dualism justifies human exploitation of all things that do not fall into the class of persons, where individual living things and whole regions of the globe may become the property of persons. Aside from the taxonomic predilection toward preloading perception with the distinction between property and not-my-property, the other consequence of propertarianism in the social sense is the undue prioritizing of property-as-value over other values, such as vital values, that are actually conditions without which property could not even exist. Propertarianism is not just the mistake of thinking that correct taxonomizing solves our ethical problems, but the mistake of privileging property over other kinds of values. Environmentalists and policy makers themselves cannot escape the structuring framework of propertarianism. Ecosystem services discourse is one example of its subtle workings. My aim in the remainder of this section is not to assess the vast literature on ecosystem services, but simply to reveal the role of social institutions and commodification in even well-intentioned environmentalist discourses.

The term *ecosystem services* is used by conservation biologists, ecologists, ecological and environmental economists, government administrators, politicians, policy makers, and increasingly the general public. Early uses of the concept intended to highlight human society's dependence on ecosystem functions and processes.[36] It implies a direct relationship between biodiversity and the ecosystem processes that sustain life. It is widely held that the number, kind, and resilience of these processes is a function of biodiversity within ecosystems, and where this diversity is reduced, the services supplied by them diminish or cease to exist.[37] Therefore, these processes have to be maintained in good working order if we would like to continue to benefit from the services they provide. Insofar as the concept is meant to refer to real ecosystem functions and processes it has an epistemological and ontological weight. When by means of it its users are able to distinguish between different types of services and rank or prioritize them (often with a view to commodifying them), it is also an evaluative, ethical and political category. As a focal point of diverse interests it is thus ripe for discussion by critical environmentalists.

The phrase *ecosystem services* (ES) entered into more common usage in the sustainability literature of the 1990s and became settled in the policy agenda in the 2000s.[38] Use of it has grown dramatically since the interdisciplinary and international teams of researchers collaborating on the Millennium Ecosystem Assessment reports chose to employ it as a central category in their framework for the assessment of global ecosystem health in 2005.[39] Its use now extends well beyond academia to reach government, nonprofit, private, and financial sectors. It has been used as a means for the economic valuation of nature, and produced the surprising result that the total value of these services was $33 trillion per year, as a conservative estimate.[40] Ecosystem services discourse has already been subject to criticism from many quarters. From classical environmental ethical perspectives, it has been accused of being one more utilitarian, anthropocentric attempt to frame nature for human use. Its frequent (but not universal) emphasis on economic valuation of services is contrasted with ethical modes of evaluation (e.g., the discourse of intrinsic value).[41] From an ecological perspective, it may be accused of an excessively narrow rendering of ecological science and understanding that lacks sufficient scientific basis.[42] From the perspective of social scientific political ecologists, it extends neoliberalist commodification into new domains.[43] While it sometimes appears sensitive to the differential impacts of environmental harms on communities and unequal responsibility for inflicting them, it still works within a global governmental policy framework which usually places planning and decision-making in the hands of those most remote from impacted communities. I think the combined weight of these critiques shows that the fundamental assumptions of the approach are inadequate to motivate and deliver social transformation. However, simply dismissing it without learning lessons from its shortcomings would be a mistake. One of the reasons for its weakness is that the discourse simply reflects dominant patterns of economistic thought, perpetuating rather than questioning the institutional frameworks within which our moral prioritization takes place. I conclude that while ES discourse may be well intentioned, given its basic assumptions it is structurally incapable of generating the changes it aspires to create.

Ecological economists Gómez-Baggethun et al. place the concept of ecosystem services into a narrative history of economics from the seventeenth century to the present.[44] They note that over the course of two centuries, the locus of economic value shifted from agricultural land with the physiocrats (from the mid-seventeenth century), to both land and labor, to labor exclusively, and then to human wants (after 1870). With the neoclassical economics of the twentieth century, land is completely taken out of account, once the notion that "natural capital" can be substituted by manufactured capital gains ground.[45] In the 1960s, *environmental* economists simply extended this logic, monetizing nature's benefits and use-values for humans by working within the standard neoclassical framework. *Ecological* economics attempts to temper this approach by insisting on strong sustainability, or the nonsubstitutability of many of nonhuman nature's goods and services, and places the economic flows of capital into the larger context of matter-energy flows in the ecosphere.

Although they acknowledge that the ecosystem services concept has not always served the purposes of commodification and that some authors have proposed ecosystem services discourse as a useful short-term tool rather than as an end in itself, Gómez-Baggethun et al. resignedly admit that these concepts are not neutral. "The spreading of the ecosystem service concept has in practice set the stage for the perception of ecosystem functions as exchange values that could be subject to monetization and sale. Hence, a controversial outcome of the economic framing of environmental concerns is the commodification of a growing number of ecosystem functions and the reproduction of market logics in the field of nature conservation."[46] In other words, propertarianism structures the discourse. For instance, ES and market approaches turn domains that were once Commons into commodities (property) accessible only to those with financial means.[47] This does not appear to be inevitable, on their account, since one can distinguish stages in the process of commodification, and they suggest that the whole discourse need not be thrown out because the last step, commodification, is a harmful one.[48] They suggest that if we make a distinction between the use-value of natural processes and their exchange-value (as Kovel also suggested), we would be able to acknowledge the services provided by ecosystems without rendering them exclusively in terms of exchange. They claim that their discursive framing as "services" does not automatically lead to commodification.[49] As conditions of commodification, institutions like private property and markets (propertarianism) have to exist in order to turn valued things into commodities. It is nevertheless clear that where such markets and institutions already exist for many goods, it is not difficult to foresee that anything suitably "prepared" for commodification through valuation procedures will be readily commodified.[50] They argue "that within the ideological, institutional and economic context in which ecosystem services science operates it is not realistic to assume that monetary valuation can be used *without acting as a driver of commodification.*"[51] Given this social institutional context,

> concepts like natural capital and ecosystem services set human-nature relations into one of utility and exchange, thereby expanding the economic rationality of the profit calculus into the sphere of ecosystems and biodiversity. Similarly, valuation methods frame choices within a narrative of scarcity, efficiency and profit, and therefore often serve as discursive framing and metrical technology for the commodification of ecosystem services.[52]

In other words, they conclude, "economic values, valuation methods and market schemes are not ideologically neutral." More subtle, internal effects of adopting these schemes include inducing a specific image of the human and human motivation, which "crowds out" conventional noneconomic and non-self-interested modes of reasoning. Entirely in line with the logic of the ecological materialist anthropology developed in the first part of the book, they argue that *Homo economicus* can be *induced* in contexts where its institutional forms (e.g., private property) and mechanisms are imported.[53] This is simply one more example of the operation of the circular processes of social cognition at work.

The effects of adopting this framework are also explored by other critical ecological economists. According to Markus Peterson et al., while ecosystem services discourse was intended to raise public awareness about human dependence on nonhuman ecological functions, the commodification of such services leads to the paradoxical effect that "ecosystem workers"—those processes and entities that perform the valued functions—are "erased" or made invisible in the process of commodification, just as human workers are alienated and made to disappear in the capitalist production of surplus value (on the Marxian account). This is an expression of the backgrounding and denial of dependence prevalent in dominant hierarchical dualism. It is worth quoting at length:

> When ecosystem services to humanity become commodities, the biotic components of ecosystems become the workforce whose labor and energy is purchased. We use the phrase *ecosystem worker* as shorthand for the organisms that produce services in an ecosystem service marketplace. Ecosystem services are offered as a proxy for the labor required to produce commodities. Thus, any discussion of an ecosystem service market should include discussion of the biota providing the service, or the ecosystem workers. The ecosystem worker sustains itself by transforming raw materials (e.g., minerals, fibers, energy, nutrients) into tissues so it can survive and replicate, thus producing a commodity desired by humanity, or an ecosystem service. As with other commodities traded in the marketplace, production logics tend to erase every trace of the ecosystem workers (biotic components of the ecosystem) and raw materials (abiotic components of the ecosystem) from the marketed commodity (an ecosystem service to humanity). This erasure of the ecosystem worker, and its replacement with money, contradicts the original purpose for reframing ecosystem functions and related biodiversity as ecosystem services, which was to increase public consciousness of the importance of biodiversity. Thus, the attempt to demonstrate human dependence on local and global ecosystems risks erasing these very ecosystems from public consciousness.[54]

Because exchange-value floats disconnected from real natural processes on such accounts, "the concept of ecosystem services has decoupled function from service sufficiently that many people may be aware of the economic value of a given ecosystem service without recognizing human dependence on local and global ecosystems."[55] This remark clearly reflects the backgrounding and denial of dependence that characterize such an approach.

These authors clearly suggest that dependence is obscured and denied by commodification. A commodity is, by definition, something unnecessary. It is "contingently valued." Of course, we can come to depend on things and services purchased, like the energy provided by the electric company, or the water provided by the municipality. But the fact that we choose to pay for them undermines the sense of their *necessity*. Where real dependence becomes invisible, prioritization is untethered from the context of real ecological relationships. Thus, markets in the essential constituents of environmental and human well-being obscure the difference between things that are fundamental, indispensable, invariable, shared, or conditioning vital values and things that are valued for convenience, efficiency, or profit.[56] Drawing on the concepts of use- and exchange-value from the Marxian tradition allows us to see the perverse prioritization at work even in these more subtle environmentalist forms. In the last section I contrast this pattern of prioritization with one more appropriate to a critical environmentalism.

POLITICAL ECOLOGICAL ETHICS

Experiential responses to long-term dependence in ontogeny—especially in western cultures—exhibit a deep resentment of dependence, along with its denial, and a damaging fixation on independence. In western cultures dependence always appears as a severe weakness, a threat to independence, an unfortunate temporary but necessary stage of development, or an episodic or chronic misfortune. The value derangement characteristic of Modernism—exaggerated emphasis on the anthropogenic origin of value in ethics and economics, denial of value to nature and life—may stem from the learned inability to affirm relations of dependence as necessary and valuable. Consumer capitalist societies built around the combined mythologies of the anthropocentric supremacy of reason and the inexorable progress and advancement of technological civilization are not psychologically or imaginatively equipped to acknowledge the embodied and embedded place of the human in nonhuman nature—its status as potential prey, for example.[57] The relentless quest for profit and material wealth allows it to deny or keep the ecological realities at arm's length temporarily, but as a long-term solution in lived duration the pattern of prioritization that correlates with this cultural model leads to catastrophe. The dominant economic social imaginary and ideology generates the self-image of a being with far more determinate impulses than it actually has—specifically egoistic, selfish impulses geared toward ruthless activity, hardwired into human nature. Systematic inculcation into productivist and consumptionist ideology produces widespread denial of dependence on nonhuman nature as well as other people, persistent self-deception and value delusions. Drawing on the account above, in this section I elaborate the contrast between two *ethe* or patterns of prioritization, consumer capitalist and broadly ecological. The process of prioritizing is the disposition to rank order values and practices in a way that privileges one value, cluster, or set of values above another cluster or set for a single project (episodic prioritization) or for the long term (chronic prioritization). The vital (systemic) values that are emphasized in the principle of survival are to be chronically prioritized.[58]

As we have observed, many environmentalists and ecophilosophers encourage us to imagine a world that is not devoid of value, or populated only with objects useful for human pursuits, but one overflowing with values of diverse kinds. The Modernist preoccupation with the subjectivity (relativity) of feeling and value preferences has served to prevent genuine inquiry into the question of the axiological dependence of values, and of their ranking. The popular conception holds that while each person may have a subjective value ranking, there can be no "objective" ranking for all cultures, genders, races, times, and places. Instead we must rely on the purely rational order of consensus, universal law, and the concept of rights (proceduralism). However, we have to ask what respect for the law or the intrinsic value of rights depends upon. It does not take much reflection to notice that the concept of a right depends on the felt value of the person, while universal law depends upon the values of equality, justice, and fairness. Rationality itself is not somehow of neutral value above it all, but is felt to be valuable as the result of centuries of struggle against superstition and fanaticism to make its claims heard. These values are themselves culturally particular; they have a history of meanings and practices behind them. But they may also be apprehended by, appreciated, and become normative for those who do not share in this particular history directly. The "same" values may very well appear in different forms among various peoples, but in different relations of priority or emphasis. This is especially the case for vital values. One ethos may clash with another not because they are entirely incommensurable, but because the same values are ranked in different orders of priority. According to the maintenance-loss model, the more we are exposed to and experience one particular pattern of prioritization, the more it seems natural and inevitable subjectively, and the more our plastic brains embody it objectively. The process of transforming the current dominant anthropocentric consumerist social ethos will be long and difficult because it is conditioned not just by ideology and the social imaginary of progress, it is embedded in institutional practices and inscribed in living brain matter. We know that to change brains and affects we need to get bodies to do different things, and to get bodies to do different things we need to change institutions; from the other end, we need to ruthlessly critique social ideologies and imaginaries in order to show the necessity of changing those institutions, which will in turn change bodies, affects, and brains.

Many environmentalists have proceeded as if reasoning with people, learning about the unintentional destructiveness of one's ways, or presenting them with information about communities who do things differently is all that it takes to produce social change. Based on the anthropological model presented in part 1 and the value theory explored in part 2, it is more than likely that social change requires being around other people with different commitments (as well as different social institutions and imaginaries) long enough to absorb and practically embody an alternative model. It may simply require living in a community with a different value array and ethos, not just hoping for one while engaging in business as usual. On the social psychological model, it is not rational persuasion that instigates change, but the triggering of a relatively small set of affective-evaluative responses in relation to new situations, objects, and relations. In such cases, deeply felt responses are associated with new perceptions, reattaching them to the different goods and situations. In this case, it may not be a matter of creating commitments from scratch, or moving from nihilism to affirmation of value, but of transvaluing (reprioritizing) what was already acknowledged to be of some value, and consistently acknowledging its priority in cases of conflict.

To give value-quality flesh to the skeletal abstract contrast invoked here, we could say that under capitalism a distinct pattern of prioritization of values takes hold of individuals that leads to the disintegration of ecosystems and persons. In terms of the value dimensions explained in the last chapter, capitalism offends against both conditioning and conditioned values at once. Processes of commodification depend on propertarianism and possession, transforming the "free gift" of nature into "natural capital." Propertarianism and expropriation are the unspoken metaethical principles of the functioning of the system. Even capitalist moral values depend on goods values, but only certain forms of them, particularly bearers conceived as property or commodities with an exchange-value. This is the institutional side of the ethos. Vital values like biodiversity and ecosystem health are anthropocentrically valued only to the extent that they are important for human production and consumption of further commodified goods and services. On the "weak sustainability" view, if the day arrives when we can technologically replicate the processes and services provided by nature, then corporations will happily replace nature with its technological substitutes. Priorities congeal around the growth imperative at the expense of all else. Profit generation is more important than human and ecological health, ecosystem integrity, conservation, biodiversity, wildness, or any other vital environmental values. Institutions are put in place to ensure that extraction of value from nature can proceed unimpeded by moral quibbles about environmental injustices or destruction of nonhuman nature. Even for the ecosystem services approach, as some authors have suggested, commodification itself effectively undermines whatever well-intended initial impulse led to the valuation of those services. On the social psychological side, success at exploitation, personal satisfaction, efficiency, convenience, and utility are taken as some of the highest values.[59] Moral value supposedly accrues to the person who pursues these values well, and this entails an at least minimal sense of autonomy or self-determination (even if tinged with self-centeredness). The allegedly descriptive axiom of economics that each agent will exclusively pursue their own interest is also morally normative; you are a better (capitalist) person if you pursue your own interest, largely indifferent to the needs of others. Haidt noted that people in capitalist societies will assess their own behavior in terms of selfish motives even when there were not any there to begin with: "Americans frequently make up self-interest explanations for their attitudes, votes, and charitable actions, even in cases where they appear to be acting against their self-interest."[60]

A critical environmentalist ethos contrasts sharply with this. Nonhuman nature's genuine agency and fecundity must be acknowledged as the ontological and axiological condition of human life and flourishing, since dependence is impossible to acknowledge if that on which we depend is not regarded ontologically as an agent or condition of our being. Its processes have to be valued as irreplaceable and nonsubstitutable. In place of the propertarian framework that classifies all goods as property or commodities, a revisionary conception of the Commons could be used to critically interrogate dominant models and may constitute an ideal to guide novel institutional initiatives.[61] Part of an adequate response to dependence involves institutional codification of the fact that nature is unowned and not the property of humankind. We live in a world not made for us. Keeping in mind the slogan that "no one owns the earth" is a helpful way to remember that the aim of sustainability will not be furthered by more private ownership of the land, air, waters, organisms, and natural processes, and will likely be endangered by it.[62] *Usufruct* has been called the principle on which Commons models of community natural resource use are based.[63] On the social psychological side, the vital values of biodiversity, fecundity, and ecosystem health may be appreciated both for their role in sustaining humans and other living things as well as in themselves. Nature-relative moral values include sufficiency, attitudes of "openness to the world," and a sense of kinship with living things. According to Kovel, in "ecocentric production," the aim is to see value in the process as well as in the goal. It finds enjoyment in labor freely chosen, not just in satisfaction by the product.[64] Pursuit of such ends leads to the recognition of moral worth in the agent's pursuit of them. Social justice, integrity, care, autonomy, respect, self-reliance, self-determination, and generosity are other moral values, practices, and dispositions which may lead to self-realization in community. It is when our autonomy is actualized, not merely in consuming to satisfy trivial interests, but in genuinely "creating ourselves" in the face of the demands of multiple values, and under substantively acknowledged embodied and embedded conditions of dependence, that we truly live.

If the environmental ethical ideal is prioritization of the conditioning as well as a synthesis of conditioning and conditioned values, then we might argue that usufruct performs such a synthesis. All humans live from nature in procuring their means of subsistence, and usufruct asks that conditioning, vital values be fulfilled fairly for all by not enclosing the Commons. At the same time, what is used ought to be improved upon, and manner and means of this improvement gives the agent the opportunity to pull this off as creatively as possible, in ways perhaps deserving of praise. Usufruct puts limits on the use of a resource, and in using it one may improve it for one's own or another's future use. Where resources are in limited supply and consumed, steps will be taken by those who have used them to replenish or restore them. This gives persons the opportunity to achieve some merit through giving of their respect, time, and care, sculpting their character in the process of provisioning the future, their own or another's. Respectfully appropriating, using, enjoying, and

Table 5.1. Two Contrasting *Ethe*

Ethos	Goods	Morals	Priority Pattern	Institutions
Capitalist	Property, exchange value, commodities, etc.	Prudence, "freedom," self-interest	Surplus value and commodities take priority over all else	Property and commodities/services disconnected from the Earth's real ecological processes
Ecological Materialist	Biodiversity, ecosystem agencies, fecundity, life, health, etc.; vital values take precedence	Openness, self-realization, community, freely associated labor, etc.; enabling values with justice on one end and self-realization on the other	Vital values take priority over exchange-value; synthesis not offending against the conditioning vital values while dynamically nourishing the conditioned	No one owns the Earth, processes and entities cannot be commodified; Commons/usufruct as model for prioritizing vital values instead of private property

improving natural resources, places, and processes differs radically from the exploitative attitudes and practices of consumer capitalism.

It might help to think of each of these briefly sketched arrays as a gestalt, the whole of which colors each individual part. But contrary to the common gestalt or holistic view, there is never total incommensurability between gestalts. There are always some shared vital values given our shared embodiment and embeddedness in ecosystems. The vital values discussed in the last chapter are the stable conditions for human reproduction, growth, and development in natural-social communities. In the last chapter I argued that there is a priority of the conditioning values. We saw that higher goods give meaning and satisfaction to existence, but no one can appreciate these (for long) if they are struggling to survive. The human and nonhuman victims of the current global system suffer from the lack of provisioning of conditioning values. Once the relations in the hegemonic capitalist priority scheme are clarified it becomes apparent that the capitalist system is designed to systematically exploit classes of victims, human and other-than-human, specifically in the way that it prioritizes exchange and surplus value over all other values, even vital values. Decisions should prioritize securing the preservation of these conditions before all else; that is, there is a certain priority of conditioning values. But this is not the only principle. Only on this basis can dynamic, forward-looking morality generate its most merit-worthy expressions, such as generosity, kindness, trust, friendship, and care for the biosphere. Stronger vital values as constituents of growth, reproduction and survival of flourishing human and nonhuman communities must be respected, given the principle of dependence, just as higher goods like knowledge systems and beautiful artwork as well as moral values have to be fostered (second principle of prioritization). These observations suggest that the aim of moral action in the midst of conflicts must be to synthesize stronger and higher values, or conservative and dynamic impulses at once (third principle). This notion reflects the way that specific kinds of moral values depend on the actualization or existence of specific kinds of vital values in an ongoing pattern of prioritization or ethos.

Transformation of an ethos is not simple and requires consciousness raising, or a critical ethical consciousness.[65] In my terms, this entails both the recognition of human and nonhuman others as autonomous centers of agency, the principle of dependence, foregrounding the problem of prioritization in the midst of conflicts, and acknowledging the need for a synthesis of conditioning and dynamic values. Community solidarity and consensus will be forged, in my view, in the course of direct democratic discussions about priorities. Value priorities are expressed in habits, practices, myths, and institutions, and all of them must become objects of inquiry, where the values are discursively articulated and their relations clarified.[66]

This chapter has shown that if they aim for significant social change, environmentalists have to take the forms of social determination at work in global capitalism seriously. However, this has to be done in the right way. A holistic account of the capitalist "force field" will have little explanatory or critical power without showing how its social determination works in different spheres. Practically speaking, holistic accounts such as Kovel's may make it harder to illuminate opportunities for intervening and engaging in social change. Regarding capitalism as "the enemy of nature" is a good rallying cry, but it makes for crude social analysis. Likewise, contrasting a relational "ecological worldview" with a demonized mechanistic one makes for good polemics, but entails an impoverished conception of motivation and an inadequate metascientific stance. Explaining why this is so in more detail requires a discussion of social and natural ontology in the last part of the book. This is because while dependence has been repeatedly invoked throughout the book, its meaning has been left somewhat undertheorized. It has been decisive for the conception of an ecological materialist anthropology of embeddedness (asymmetrical dependence), as well as for the conception that some values depend on others as well as on bearers of value. Engaging in some further ontological discussion is necessary to explain why understanding dependence in the right way is important for the ecological materialist critical environmental philosophy developed here, as well as to explain why this position itself should not be understood as a new "ecological worldview" or as some version of science-driven environmentalism. Therefore, a detailed treatment of the concept of dependence as it informs anthropology, value theory, and an adequate metascientific stance for environmentalism will be covered the last part of the book.

PART III

Ecological Ontology

CHAPTER SIX

Metascientific Stances and Dependence

ENVIRONMENTALISM AND METASCIENTIFIC STANCES

The classical problems of environmental philosophy already discussed, anthropocentrism and intrinsic value, were often accompanied by claims that in order to transform an ecocidal civilization it is necessary to adopt an ecological worldview. Such a worldview would answer questions not just about the place of human beings in nature and the value of nonhuman nature, but also about the nature and role of the sciences in relation to environmentalism. On an ecological worldview—where some form of organicism, holism, or relationalism are promoted—humans are to be regarded as part of nature and nature is to be regarded as intrinsically valuable, thereby providing grounds for resistance to exploitative anthropocentrism. This ecological worldview was taken to be quite different from that of a reductionist, mechanistic science, and the model of "worldview clash" between broadly ecological and mechanistic perspectives can be understood as a type of metascientific stance that not only frames the way that the sciences are considered by environmentalists, but also functions to provide a kind of global orientation to ethical, political and anthropological questions.

By *metascientific stance* I mean implicit conceptions of the role and place of the sciences in the production of knowledge (including notions of epistemic authority, social context, historical significance, and political relevance). In other words, a metascientific stance includes tacit assumptions about the nature, practices, goals, and place of the sciences in society. Although a metascientific stance is not identical with an implicit or explicit ontology, such stances entail many ontological and epistemological claims. The global orientation such stances provide often speaks to basic ontological assumptions about agency, determination, and dependence. In this final chapter, I examine the way that the principle of dependence invoked throughout the book can inform a more nuanced metascientific stance that recognizes natural and social dependencies in their multiple forms, and I also look more closely at the conception of dependence that this and other stances imply. For instance, what I call the "worldview clash" and the "science-driven" environmentalist models obscure dependence in different ways. Worldview clash does so, first, by generally regarding dependence as interrelation in its ecologically informed view, which fails to recognize the ever-present structural asymmetry of material dependence (and often entails antirealist epistemological assumptions). Relatedly, science-driven environmentalism—while it apparently preserves some idea of human dependence on nature, as in, for example, ecosystem services discourse—systematically obscures the social dependence of scientists and other knowledge makers and the heterogeneous conditions that contribute to knowledge-making in social contexts. Both models, I conclude, misunderstand natural and social dependencies ontologically. A better metascientific stance allows us to avoid the pitfalls of both of these options. While much ontological reflection is called for here, I only include enough of this for the reader to appreciate why traditional environmentalist understandings of the place of the sciences in society are also shaped by dualisms and mask dependence, and so must also be set aside in a critical environmental philosophy. More sophisticated ways of rendering science-environmentalism relations are needed in order to understand knowledge production in social contexts as well as the potential for intervention in them. As in anthropology and value theory, this too calls for an understanding of the asymmetrical dependencies in the structure of the real world. Political ecology, or critical environmental philosophy, cannot do without this conception of asymmetrical dependence in anthropology, ethics, or metascientific stance.

    Helen Longino convincingly argued in the early 1990s that "if no project of political transformation in the twentieth century can do without science, then neither can we do without a better philosophical understanding of scientific inquiry than is currently available."[1] This claim has lost none of its urgency for the twenty-first century, and environmentalism as a multifront project of political and cultural transformation is no exception. The philosophical understanding of scientific inquiry of many environmental philosophers and ethicists is generally of the worldview clash or science-driven type. If popular metascientific stances are inadequate, what are the alternatives? From the start, most environmentalist practicing scientists (such as Barry Commoner) did not embrace the worldview clash model perhaps because it seemed to be outright anti-science, or because it makes invisible many of "the possibilities of modifying scientific practices better to correspond to the demands of an egalitarian society and better to cope with ecological issues."[2] While we should not demonize the sciences, we also should not be too sanguine about the possibilities of modifying scientific practices, of course, and should seriously "wrestle with the potential and limitations of critical reflection as a means to redirect practice."[3] These remarks from a few politically savvy scientific ecologists already indicate more reflexivity about the practice of science than is usually acknowledged in popular understandings of the sciences and science-driven environmentalism. The popular understanding—embraced by practitioners, policymakers, and the public—relies on a conceptually bankrupt positivistic model of science that assumes a sharp boundary between observation and theory, a radical disjunction between subjective and objective constituents of knowledge, the unity of the sciences ideal, reductionist metaphysical materialism, and a gradualist model of scientific advance. The problems with this model have been discussed for decades in mainstream philosophy of science even though it still persists in popular consciousness and also informs the science-driven environmentalist's metascientific stance.

In contrast to both of these models, a stance is needed that advocates an account of social knowledge production that rejects the stereotypical rational-social dualism in accounts of the sciences, and leaves ample room for discussion of both the role of values as well as of materials, practices, traditions, and institutions in knowledge-making. On dualistic conceptions, science is either the product of reason mirroring nature or the product of communal social interests. This exclusive opposition is clearly just one more version of the human-nature dualism we have been dealing with throughout the book.[4] We can recognize the dualism in the fact that for both the clash of worldviews and science-driven environmentalist models, it is what uniform Nature is like that is the reference point for recommendations of uniform (Human) social or behavioral change. They obscure the yawning gap between theory and practice with assurances that recommendations for social change follow quite directly from "the way things are."[5] Both equally rely on the conception of an undifferentiated Humanity with an identical interest in escaping environmental degradation, and both downplay the important role that real social institutions and intersecting processes play in the construction of knowledge (of Nature). Avoiding this dualism means understanding scientific work in its social context, as well as the historical, social, and philosophical interpretation of that work, and has to be part of answering the question about how ecology is used or may be used to support environmental philosophy.[6]

These conceptions have practical consequences. Both worldview and science-driven conceptions take substantial agency away from those in communities of struggle, the first by making environmentalism a matter of worldview conversion usually without reference to social institutions, and the second by making it the business of an elite policymaking class. Environmentalist knowledge-making has to be seen as another important aspect of political ecological labor. A pluralist metascientific stance contrasts with both worldview clash and science-driven environmentalism in rejecting the universalist thesis (that there is a uniform human group and homogeneous human concern with avoiding environmental degradation), and adopts a multifactoral approach to considering science-politics relations. It makes visible the multiple opportunities for intervening in ongoing social processes at local scales that are rendered invisible by popular worldview science-skepticism or positivist science-driven environmentalism. It allows us to satisfy our need for local explanatory efficacy as well as for a global structural account that may orient individuals and communities endeavoring to thrive in a more-than-human world, a world not made for us.

The ecological materialist stance adopted here is not another version of the ecological worldview, nor does it accept science-driven environmentalism's positivist conception of science and action. Its metascientific stance is informed by a specific ontological stance that has been implicitly relied on throughout the book. Firstly, we have seen that in philosophical anthropology, we do not have to choose between nature or culture—our "natural artificiality" means embodiment and plasticity scaffold cultural achievements prepared for in the womb. These cultural achievements, including ecological ethics and politics, are not explicable in terms of the usual evolutionary explanatory strategies. They are scaffolded by and diffusely dependent on vital human capacities and the whole more-than-human world. Secondly, in value theory, these cultural categories include the categories of goods and moral values. Vital capacities provide the naturalistic condition for the existence of vital and moral values (in humans), and moral values are themselves both superposed on and intensionally included in goods values in characteristic relations of dependence. Finally, a choice of metascientific stance takes place under material conditions as well, where the social conditions of history and culture are scaffolded by the vital capacities of agents as well as by the structures of the vital-physical world. In all of these dimensions of human experience, there is a macroscopic ontological order of structural dependence that can be called a relation of "stratification." This conception of stratification is what gives coherence to all of the senses of dependence discussed above, and provides an orienting perspective on each of these problems. I explain this understanding of stratification in the third section of this chapter.

The next section describes an intellectual trend in science interpretation that moves from a dualistic view, to a more reflexive model, to the recognition of complexity, and finally to dependence. Once we make the "social turn" in our interpretation of the sciences, we can no longer entertain the positivist idea that science gives us the decontextualized facts and laws that philosophers and policymakers may then use to justify their recommendations for human-nature relations. But neither can we endorse a generalized constructivism that anthropocentrically foregrounds the role of human minds in shaping the real world. Once this dualism has been undermined, a reflexive view of the "unruly complexity" of knowledge production is introduced that provides the best relational ontology for science studies that accounts for the multiple factors involved in the scientific production of knowledge. Even this relational view, while better than most, still makes it impossible to recognize the asymmetrical principle of dependence defined in this book. Once we see the limitations of the best relational ontology for science studies available, we can better understand the justification for adopting a stratified ontology. I argue that an account of ontological stratification ensures that asymmetrical dependence will no longer be obscured by reductivist explanatory strategies, either from the direction of nature or of society, nor even obscured by relational ("flat") ontologies. The next section prepares the way for the adoption of a pluralist, political ecological, metascientific stance that can embrace the conception of science as social knowledge as well as an ontological realist principle of dependence that provides global orientation for critical environmentalism.

FROM RATIONAL-SOCIAL DUALISM TO UNRULY COMPLEXITY

In the fields of philosophy of science and science studies, scientific knowledge is primarily interpreted either as the product of rational, cognitive operations on evidence and statements, or as the result of social practices, ideologies, or interests. It is almost never interpreted as both rationally and socially produced at the same time. This contrast manifests another typical dualism in Plumwood's sense, where the social explanation is inferiorized by philosophers, while conversely the rationalist explanation is inferiorized by sociologists (broadly construed, including historians, anthropologists, science studies theorists, etc.). Helen Longino examined the arguments of both camps, contended that both groups implicitly adopt a dualistic construal of the rational and the social, and showed how each unjustly minimizes or backgrounds the claims and resources of the other. What Longino herself was after was a reconciliation of the two, an interpretation that gave both the social and the cognitive a role in knowledge-making. She aimed to "offer an account of scientific knowledge that not only avoids the dichotomy but integrates the conceptual and normative concerns of philosophers with the descriptive work of the sociologists and historians."[7] She undermines the dualism by decomposing the associated dualisms that support it. Stereotypically, positivist epistemology assumes the rational integrity of the individual knower, the metaphysical unity of nature, and a firm distinction between the context of justification (objective) and context of discovery (subjective). In contrast, sociological accounts assume social knowers determined by context, the metaphysical plurality of nature and society, and the blurring of the justification-discovery distinction. Dualizing philosophical thought places the first cluster under the rubric of the rational, and the second under the rubric of the social. Longino nicely employs Plumwood's strategy of demonstrating continuity rather than radical exclusion by dehomogenizing within these broad classes. For example, if a rational agent insulated from the contamination of the social is presupposed in traditional epistemology, and if for holists the individual and her beliefs is instead taken to be a product of social processes, these exclusivist views can be mitigated by one that minimally acknowledges the interdependence of individual knowing agents with their social surroundings. Likewise, against pure constructivism about objects of knowledge (facts are made) and traditional scientific realism (facts are just there to be discovered), she proposes that no single situated account or model captures the entirety of natural-social processes, but each does capture some aspect of reality.[8] Finally, in place of the ideas that context is irrelevant to justification and its opposite (all accounts are subjective), she substitutes a "contextualism" where "justification is neither arbitrary nor subjective, but is dependent on rules and procedures immanent in the context of inquiry." What these internal distinctions importantly provide is a way to break out of the apparent self-evidence of the original rational-social dualism.[9] Although I cannot fully examine her response here, her work should be viewed as an important resource for generating a pluralistic, critical metascientific stance for environmentalists.

Philosophers and sociologists are not the only ones to have been troubled by traditional dualistic conceptions in the sciences. It is important to note that some efforts have been made by ecological scientists themselves to move beyond this dualism, even if not as explicitly. This trend is evident in some discussions of paradigms in ecology and of the nature of reflexivity (the idea that categories used to study objects of inquiry must also be applied to researchers themselves). Ecology is usually defined as the study of patterns in nature emerging from the relations between living things and their surroundings. The economy of this formulation masks significant debate about what the fundamental objects of inquiry in ecological science and its subfields are, whether general laws can be established regarding these objects, and even over the scientificity of ecology as such. These persistent debates have not been resolved, but only compounded, by new theoretical developments in the field. At the start, it must be acknowledged that ecology is not homogeneous, and onlookers and practitioners alike have to recognize the differences between at least two major approaches within scientific ecology, population-community ecology and ecosystem ecology, which may be considered the two major competing paradigms within ecological science.[10] Put simply, population ecology often seeks to explain patterns and causes of change in the distribution and abundance of organisms in space and time, where abiotic factors are taken to be external to the relations among organisms, and the ecosystem remains a static backdrop to the adventures of individuals within them. It has an individualist ontological cast. Systems ecology links the patterns of energy flow through ecosystems to the trophic physiology of organisms themselves. The abiotic environment is explicitly included, and heterogeneous individual organisms are often lost sight of for the sake of mapping materials and energy flows. It is ontologically holist. Both major paradigms are continually contested and under revision.[11] The variety within ecological sciences themselves could be further explored, but this is enough to frame further theoretical refinements introduced by some ecologists. These refinements involve a conception of reflexivity that begins to break down the dualism shaping science interpretation.

O'Neill et al. aimed to resolve some of the tension between the population and ecosystem paradigms by claiming that both sides may be justified in their choice of primary categories provided these categories are not considered absolute, but relative to what they call "observation sets" and considerations of ecological scale.[12] An observation set is "a particular way of viewing the natural world. It includes the phenomena of interest, the specific measurements taken, and the techniques used to analyze the data."[13] Spatial and temporal scale are important components of the set, and since scales vary and different points of view emphasize different measurements and phenomena, no single perspective can provide a comprehensive explanation of the system. Consequently, they argue that it is "impossible to designate *the* components of *the* ecosystem."[14] One artificially limits theory by building it on the basis of a single perspective. The point here is that by employing the language of observation sets, analytical categories are made relative to epistemic perspective. Whatever its methodological value for ecology, this model manifests an epistemological stance that accepts an irreducible relation between scientific investigators and their objects of inquiry. That is to say, it adds a reflexive dimension to ecological epistemology. This is one important aspect of the "picture of ecological complexity" that has emerged in ecology since the 1980s.[15] Note that the observation set as a way of viewing the world includes the researcher, object of inquiry, tools for measuring, and measurements, but as yet makes no reference to social context. The introduction of reflexivity does, however, begin to undermine the rational-social dualism enforced by positivism.

In another more recent pioneering work of integration, Pickett, Kolasa, and Jones further elaborate the individualist and holist paradigms in ecology. Like O'Neill et al., the authors make the point that the individualist and holist paradigms have seldom been explicitly recognized as such by practicing ecologists, and that much confusion has arisen from the misapplication of categories at home in one paradigm to another. For Pickett, Kolasa, and Jones, the observation sets referred to above are taken to be products of an implicit paradigm. While the distinction between observation sets and paradigms seems to be epistemologically negligible, the step from observation sets to paradigms is in fact a significant one since the concept of *paradigm* explicitly acknowledges the historical and social aspects of scientific theory and practice in a way that *observation set* does not.[16] They draw attention to the fact that different theoretical and methodological emphases may correspond to different coexisting paradigms. They define a paradigm as "*the worldview, belief system, series of assumptions and techniques, and exemplars for problem solution held in common by a scientific community.*"[17] In history and philosophy of science, adopting the paradigm concept normally implies a rejection of strictly positivistic approaches to the sciences, and entails the recognition that the practice of science is historically and socially conditioned, at least to the extent that scientific knowledge is produced within a certain community of scientists. These authors emphasize that the term does not imply the total relativity of knowledge since paradigms are intersubjectively shared by community members and form the common conceptual framework in terms of which new projects are developed, questions are asked and answered, and experiments are carried out. While there are clearly subjective elements involved in choice of methodologies or in background assumptions, objectivity can nevertheless be achieved provided such social knowledge passes through processes of conceptual criticism.[18] These authors are interested in unifying the ecological sciences, and in integrating the various paradigms in order to increase ecological understanding. Whatever the case with these aims, or whether many ecologists do or would subscribe to their project, the point of drawing attention to this trend in theoretical ecology is to show that adopting the paradigm framework amounts to a new metascientific conception of the relation between knowledge production, social context, and social engagement. While this goes further than its precursors to encourage reflexivity and break down the rational-social dualism, due to its holistic epistemological assumptions, however, it still does not pay enough attention to noncognitive institutional, material, or practical factors in knowledge-making. Funding might push research down one path rather than another, or easy access to some computer-modeling software might make one course of action more likely. Although the introduction of greater reflexivity into the approaches of some ecologists has initiated a breakdown of traditional epistemological dualisms, the work of some political ecologists makes even more room for considering the roles of material factors in knowledge production.

Ecologist and science studies theorist Peter Taylor also aimed to undermine a number of long-standing dualisms, including that between the rational and the social in accounts of scientific practices. A former student of Richard Levins, Taylor examines the significance of complexity in ecology and socio-environmental change, where interacting social agents "establish what counts as knowledge." In his work he attempts to foster interpretations of these interactions that will influence ongoing interactions or "link knowledge-making, interpretation, and engagement in social change."[19] His approach builds in a robust reflexive dimension that effectively challenges persistent dualisms. One of his aims is to encourage all researchers to become more reflectively self-aware in dealing with complex situations, and to recognize the limitations of such reflection in redirecting ongoing processes. They should take steps "to reconstruct the unruliness of complexity without suppressing it, to link knowledge-making to social change, and to wrestle with the potential and limitations of critical reflection as a means to redirect practice."[20] In several analyses of ecological and social networks he undermines the dualisms agency-structure, mind-body, human-nature, external-internal, realism-constructionism, objective-subjective, and knowledge-action (or theory-practice). Taylor adopts a metascientific stance that makes room for diverse factors in the production of knowledge, without privileging any single element. It will be beneficial to spend some time exploring a few ideas of Taylor's before moving on.

Taylor resists any approach that—through its assumptions, practices, and categories—*prematurely* imposes too much order upon unruly complexity. In terms of ecological theory, both population-community and ecosystem-process models are guilty of this. Taylor claims that both approaches "should not assume that ecological complexity can be partitioned into communities or systems that have clearly defined boundaries, coherent internal dynamics, and simply mediated relations with their external context."[21] Correspondingly, he advises the same caution in social studies of the sciences, where any "macrodetermination" of science's content by "capitalism" or "instrumental rationality" already prematurely organizes unruly social and natural complexity into ready-made compartments or chunks, with clear vectors of determination assumed to exist between them. To avoid imposing too much order prematurely, he recommends some guiding principles for analysis. Taylor's work is incredibly rich in insights and detailed case studies, but I will restrict my presentation to three of its aspects that will allow me to provide a general sketch of the advantages of the approach. These are its (1) ontological pluralism and parity of factors in place of dualism or monism; (2) its social conception of (human) agency; and (3) its emphasis on material practices in addition to mental or social factors. These features are evident in his specific accounts of scientific practices and social-environmental change, and I suggest that they provide crucial elements contributing to an appropriate metascientific stance for critical environmentalists. Ultimately, while Taylor's approach is best of its class, it does not allow us to recognize relations of diffuse dependence, which also means that it does not readily provide a globally orienting function for critical environmentalism.

(1) Ontological pluralism is meant to resist characterization of complex situations in terms of ready-made dichotomies, like the rational and the social. We ought not assume in advance that knowledge is "determined" by the society or context in which it is made, or by the nature of the "real world" over

against society—it is "heterogeneously constructed" from many different features.[22] This means that single-dynamic accounts of knowledge production, such as "interests determine knowledge" or "the data speak for themselves" are hopelessly simplistic.[23] What he calls "heterogeneous webs" of factors include various components of the physical and social world, contingent on the situation in question, but possibly including soil, bodies, farmers, books, computer hardware and software, spatial distances, scientists' or investigators' assumptions, economists, objectives or goals, and so on. There are no a priori reasons to exclude any of this from knowledge-making networks. A key premise is that all of these factors stand in ontological parity.[24] Such accounts should also display an "intermediate complexity" which allows one to avoid macrolevel dualized oppositions, such as "ecology" versus "economic rationality," as well as grand narrative transitions, such as "from feudalism to industrialism," or "organicism to mechanism."[25] Importantly, there are clear social-political consequences of adopting such an account. For example, if accounts of environmental degradation must be complex, and must take stock of the wide range of factors involved in generating it, then the responses to it will likely be just as complex. For example, if there is diffuse and distributed agency in the causes of global climate change, then responses that focus only on one type of agent or factor are bound to be inadequate. One significant practical upshot derived from this pluralistic perspective is that "intersecting processes accounts do not support government or social movement policies based on simple themes, such as economic modernization by market liberalization, sustainable development through promotion of traditional agricultural practices, or mass mobilization to overthrow capitalism."[26] This implies that worldview conversion and positivist science-driven environmentalism are also inadequate responses to ecological crisis.

(2) Taylor also resists conventional ways of thinking about human agency and social structure in social theory. In social constructionist accounts, knowledge makers "can be seen as ciphers for society or dupes for interests," and in positivist scientific accounts, agents can be forgotten "once they have helped to establish the knowledge."[27] These two interpretations of agency are another manifestation of the rational-social dichotomy. On his account, "there is no reduction to macro- or structural determination" as in the Strong Programme of the sociology of scientific knowledge, "nor is the focus on transactions among concentrated individual agents," as in positivist accounts.[28] Scientists are "practically imaginative agents" and simultaneously "represent-engage."[29] What this means on his account is that for the most part, scientists make models of

natural processes, adhering to some community standards or norms of conformation, but they also seek cross-reinforcement through analogies, metaphors, diagrams, and other social factors. If we ask "how much of the model was dictated by nature and how much by society," Taylor explains, we have missed the point of this kind of analysis. Taylor himself uses the term *construction* to designate a real process of linking components serially in time and laterally in space, where agency is "distributed" across many nodes, and this acknowledges that it is linkages and relations between components that are themselves agencies of a sort, rather than reserving the idea of agency to "concentrated" causes alone. Because of this, it is difficult if not impossible to assign responsibility for some result to one type of cause rather than another, since there are many paths to the same result, and many linkages are involved in multiple constructions that are ongoing. There is no "Nature" to be contrasted with "Society." In his terms, there are heterogeneous webs of resources where causes can be identified as contributing to heterogeneous constructive, intersecting processes, but are not "partitionable" into wholly "natural" or wholly "social" classes of resources. He resists moves to homogenize or simplify agency, and retains a notion of the imaginative, creative action of human agents without unduly privileging the mental over the material-practical. Agency is "distributed," in the sense that agents are not able to do much at all without building connections among resources.[30] This take on agency also allows us to consider values to be one set of resources or constraints among others, not unduly to be privileged in the account, but not dismissed either. Epistemic values and practical priorities both play a role in knowledge-making. Tacitly affirming a value pluralist approach, Taylor appropriately contests the idea that any agent ever pursues only one goal in any engagement, "whether that goal is revealing the nature of some underlying reality (truth), establishing instrumentally reliable knowledge . . . , furthering the interests of the agent's social group, or maximizing and concentrating the agent's social resources."[31] Plural and potentially conflicting value-driven motivation is the norm, not the exception.

(3) A final example of Taylor's undermining of dualisms is the way he shifts the emphasis away from explanation in terms of mental entities and toward "practices." This is where we can see the sophistication of Taylor's account in contrast to the other reflexive approaches mentioned above. Acknowledging work on the role of metaphors in science, he goes a step further to question some "meta-metaphors" that frame these theories themselves. These include the assumptions that "mental things—thoughts, expectations, images we have seen—shape our actions," and that "culture or society gets into these thoughts,

expectations, and images (and thus we can be taught how to perceive the world)."[32] The meta-metaphors of "culture inside thoughts" and "mental shaping action" are central to macrodetermination, worldview and paradigm-type approaches in environmentalism. The significance of this insight for the argument of this chapter can now be restated: the worldview clash model is based on the assumption that the mechanistic worldview shapes or determines exploitation of nature, while an organismic worldview would ostensibly have better practical consequences. This very assumption is questionable. It relies on an untenable holistic epistemological assumption (namely, that perception is completely theory laden) as well as an untenable conception of bridging the reflection-action gap. In addition, a crude environmentalist narrative of a new historical transition from mechanism to organicism reinforces the idea of the primacy of the mental, or the idea that "understanding the world is about living inside stories." Instead, Taylor's approach "relies less on images of mind-based subjectivity . . . and attempts to make more space for examining the material aspects of specific scientific practice."[33] He strives to displace the emphasis on these meta-metaphors by holding that "action and thought are jointly shaped through *practical activity* in which a diversity of resources are employed."[34] In other words, while it is clearly anthropologically warranted to make much of our language-and-culture-saturated mode of life, it cannot be to the exclusion of embodiment and embeddedness in practical and material networks of exchange, linkage, reproduction, and development. This view is entirely compatible with the ecological materialist conception of human engagement in the world outlined in part 1. Taylor himself also helpfully identifies unacceptable assumptions in the popular science-driven environmentalist stance as well. In many interesting analyses, Taylor shows how heterogenous webs, including global background assumptions and material resources, produce knowledge about "global environmental problems." For example, he discusses the assumptions involved in the science-politics of the Club of Rome's *Limits to Growth* study of 1972, and how many of these are still unknowingly adopted by contemporary writers.[35] He reveals that while they are well intentioned, science-driven environmentalist analyses are poorly equipped to reflect on and modify their metascientific stance in ways that would lead to local self-determination in responding to ecological crisis in place of universalizing policies. These are some of the most insightful and challenging discussions of environmental knowledge production out there, and readers would benefit from acquaintance with them. This quick look at a few of Taylor's insights gives us the resources we need to move on.

Before moving to the next section to reveal the ontological limitations of the relational approach, I will note that the value theory developed in part 2 can and should be included in this metascientific stance. Both Longino's and Taylor's proposals provide far more nuanced accounts of knowledge production than existing environmentalist metascientific stances, and both implicitly invoke the need for reflection on the goals of knowledge production projects. As argued throughout the book, without values terms we have no way of expressing different qualities of situations that we would like to bring about or avoid. Goals imply values, and being clearer about values allows agents to be clearer about goals and motives for pursuing courses of action. What makes one account of nature better than another has to be articulated in terms of some epistemic, social, and practical values, since we have no other way of articulating how and why some accounts are better than others. What leads political ecology to construct better knowledge or accounts of environmental-social problems than universalizing worldview or science-driven approaches? A political ecology is a better approach than science-driven environmentalism because, among other things, it includes the practical values of autonomy, self-determination, and self-organization for globally differentiated communities in their responses to environmental disruption. By probing the dimensions of value, practical activity, and material institutional resources in human experience of social-environmental problems, we would be better placed to deliberate about value priorities, and to reflexively inform ongoing processes of social change involving knowledge-making.

That is one crucial feature of a novel critical metascientific stance for environmentalism. We have to acknowledge that these value-oriented environmentalist projects are incomplete without supplementation by a theory of value that fills in the picture of value pluralism for the purposes of framing the practical goals of research projects and prioritization. For example, Longino's project for a feminist science involves "social values management," where (at minimum) the values of egalitarianism and liberation are weighed against the disvalues of inequality and oppression. As many ecofeminists have argued, there are direct historical and conceptual linkages between the struggles for women's liberation and nonhuman animals' or nature's liberation. One of these is the way that human liberation often involves uninhibited opportunities to live from nonhuman nature; since vital values can only be borne by specific kinds of bearers, and existence of vital values is necessary for the realization of many moral values (including self-determination), the continued existence of "wild" nonhuman nature is necessary for the realization of moral values essential to feminist and other liberation projects. Taylor's political ecology aims to bring interpretations of scientific practices and knowledge production to bear on ongoing research. Such interpretive interventions are also necessarily, tacitly or not, guided by values. He agrees with Longino when he says that "researchers need not wait, then, for the inevitable distortion of results or for the eventual acceptance of the truth; they can instead attempt to change the particular aspects over which they have the most influence."[36] Researchers interested in intervening in ongoing interacting processes must be no less guided by specific articulable values that are prioritized over others, and articulating them in social communicative action is an important step toward redirecting these processes for socially engaged agents (even if it is not the only mode of intervention). Thus, the axiology and taxonomy of values developed in part 2 should also be incorporated into the metascientific framework by helping to articulate the epistemic values and practical goals discussed above. Conjoining these analyses with the "political ecological ethics" of the previous chapter and its claims regarding a transformative ethos in solidarity and resistance to capitalism would be a fruitful endeavor, but cannot be pursued further here.

I began this section by appealing to Longino's critique of the rational-social dualism, showed how the concept of reflexivity in scientific ecology resonated with it, introduced the ways that Taylor took reflexivity and critique of dualisms in knowledge-making even further. This kind of political ecology should inform our conception of science-environmentalism relations and be used to resist traditional worldview clash and science-driven approaches. However, although this pluralist conception of knowledge-making has clear advantages over other metascientific stances, like them it tends to minimize the question of dependence. While it recognizes the dependence of scientific knowledge production on a host of social and material factors, the ontological parity it endorses tends to obscure the large-scale relations of structural dependence that have been invoked throughout the book. In other words, an ontology that appreciates the relational parity between elements is important for ecology and for analyzing social knowledge production, but it also masks macroscale structural dependencies behind a blanket claim of all-pervasive interrelations, networks, or webs. Thus, a stratified ontology has to support and supplement this political ecology of unruly complexity by further refining the sets of categories on which multifactoral accounts draw, and by preserving the principle of dependence in the structural order of these categorial principles. It serves the global orienting function currently filled by worldview and science-driven approaches by decisively foregrounding the principle of asymmetrical diffuse dependence.

STRATIFICATION AND DEPENDENCE

The pluralist approaches espoused by many of the writers discussed throughout this book help to dissolve key problems in environmental philosophy by recognizing that they stem from dualistic construals of basic categories. In philosophical anthropology, they reveal the poverty of both biological determinist as well as idealist, human exceptionalist accounts of human nature, making room for a nonreductive naturalist account of intermediate complexity that remains in contact with experience and forms an alternative metaethical foundation for environmental ethics. In value theory, pluralism allows us to avoid pernicious dualisms in theory (such as reason-emotion and needs-wants), and to circumscribe the domain of values in a way that finds them irreducible to any more basic units of ethical discourse. Opening up the domain of values in a pluralist way is indispensable for coming to terms with the pervasive problem of priorities—one that we face everywhere we experience conflicts of values—in a manner that reveals the need for cooperative communal conversation about them. Finally, pluralism makes possible a metascientific stance for environmentalists that avoids the rational-social dualism in interpretations of the sciences, and considers the widest range of factors to be potential resources for the production of environmental knowledge in social-natural contexts. This last form of pluralism, or the basic principle of "ontological parity," is an important constituent of political ecology, but it requires supplementation by a theory of stratification that recognizes the regular macroscale relations that structure the real world in which social-environmental problems occur. The conceptions of embeddedness in philosophical anthropology, of conditionedness of some values by others, and of knowledge production by material factors rely on an ontological sense of dependence that is not simply causal. The layering of social knowledge-making processes or moral agency on natural and material processes is not explicable in terms of the trite idea that everything is interrelated (by using a single set of explanatory categories), because the categories we use to capture the realities of social and moral experience, psychological motivation, vital ecosystem processes, and physical dynamics are not reducible to a single set. Here too I can only provide a summary conception of stratified ontology, and will leave a more detailed account for another occasion. The point of introducing it, even briefly, is to show that it can provide the global orienting function currently provided by inadequate worldview or science-driven approaches.

According to Taylor's relational approach to science studies, heterogeneous components are linked in complex networks of relationships that lead to the production of social-ecological knowledge. In order to avoid a kind of categorial reductionism, however, the heterogeneity of factors relative to one another must be preserved and not reduced to a single set of explanatory categories. In the ontological structure of the world, the real dependence of living matter on material regularities, of the mental on the living, and of the social on the mental, is structurally invariable.[37] Ontological pluralists must resist the reduction of categorial plurality as well, and conceive of the categories belonging to specific domains of phenomena, such as physical things, living matter, mental processes, and social life, as coherent sets or clusters of categories with internal relations among themselves and relations of stratification between the domains of phenomena they describe. This is very different from the more conventional granular hierarchy of levels, such as the scalar hierarchy of dynamical systems, ranging from the microphysical domain of elementary particles to the macrophysical realm of star systems and galaxies. Forms in such a scalar series are considered primarily in terms of the fundamental ontological categories part-whole, matter-form, and element-system. All of these pairs are relative in the sense that at one level a cell, for instance, may be a whole while at another it forms only a part, or it may be regarded as an element of a system at one level, or the system itself at another. Cells are one kind of concretum to be explained by the appropriate set of categories, and clusters of categories maintain coherent relations with one another. These categorial relations are different from those that concreta have with one another.[38] Stratification of categories in this sense entails the irreducibility of one set to another. The structural macroscale dependencies in the real material world *are* this very stratification, this superposition of categorial principles upon principles. Explaining how this works requires a quick digression on the topic of categories before returning to the connection between metascientific stance, dependence, and global orientation for critical environmental philosophy.

Categories ought to be regarded as internally complex, substantive principles immanently determining the concreta that we seek to explain. Two broad classes of categories should be distinguished. "Generic" categories run through all the concreta of the world. There are none more basic, in the sense that explanation always ultimately resorts to categories of this sort.[39] Generic categories include those such as principle-concretum, form-matter, unity-multiplicity, substrate-relation, and element-system. The pair determination and dependence is implicated in the principle-concretum pair (concreta are *determined* by principles, the latter *depend* on concreta), as are unity-multiplicity (the same *one* principle determines *many* concreta of the same type), and element-system (every stratum of categories forms a coherent *system* of which each is an *element*). Counter to the many flavors of monist and dualist frameworks in our tradition, this set of generic categories is relatively large and highly complex, a multidimensional network of mutually implicating pairs that structures the world of concrete contents. While these generic categories are universal, their universal scope is paid for at the expense of substantive content. They apply everywhere, but that leaves them relatively empty. Specific categories are more contentful, and as peculiar to particular types of concreta they are not everywhere applicable. Categories may belong to all the strata of the real world (e.g., form-matter), some of them (e.g., spatiality), or just one (e.g., purposive action). Only the explanatory combination of these general and special categories brings us to the conception of singular, fully determinate, unique individuals, the concreta that populate the real world. Let me talk about the distinction between determination and dependence in order to forestall any misunderstandings that may arise.

Determination and dependence are equally fundamental, contrasting categories. The genetic bias of Modern ontology tends to conflate dependence and determination, as well as radically limit the forms of determination. Mechanism emphasizes almost exclusively locomotive causality in the physical domain, for instance. But it is evident that determination and dependence are not the same. As we observe phenomena in the world, we generally recognize dependence first, and then become curious to understand the specific type of relation between condition and conditioned. We know that the tree *depends* on the seed somehow, but only with great effort do we understand how the seed *determines* the tree (and the historical progress of this understanding is still incomplete). We can say that minds *depend* upon bodies, but not that they are *determined* by bodies without further analysis that tames the unruly complexity of processes of determination or causation to a few identifiable types. Humans depend on the kind of atmosphere, food, bacteria, and gravity provided by the planet Earth, not those of some other planet. This does not mean that the food we eat, air we breathe, bacteria that inhabit our gut, or gravity that daily constrains and enables our action determine us to behave as we do; but we do not lead a life in developmental duration without them. Additionally, wherever specific processes of determination are revealed to exist in the relations between tree and seed, mind and body, or human and nonhuman nature, the dependence relation also still exists and cannot be explained away. Thus, at least three features of dependence should be kept in mind:

(1) These examples show that dependence implies a kind of *duration* or temporal extension embracing growth and reproduction of the being(s) in question over the long term rather than in some limited segment of time or time point.

(2) For some types of dependence, the relation is always asymmetrical. In such cases condition and conditioned are not "interdependent," "reciprocally dependent," or "interrelated," as relational ontologies often emphasize. This means that for ontologies that are built on their opposition to substance metaphysics and instead privilege relationality and interconnectedness, it is not sufficient to be indiscriminate about the relations that are emphasized. Some relations are more important than others for some reasons. For the project of environmentalism, the relation of dependence is of primary significance and is not reducible to just another "interrelation." In order to get out from under the compelling but only minimally effective emphasis on "relationalism" it is helpful to consider a case where dependence can be clearly distinguished from other kinds of relations.

In the history of epidemiology, a correlation between mosquitos and malaria was all that was needed in order to engage in action to stop the spread of the disease. Researchers did not need to know the precise causal mechanisms involved in transmission before action was possible. A correlational dependence between the presence of mosquitos and resulting malaria was enough to act on; mechanisms of determination came later and allowed us to refine prescriptions for action, but are not necessary for them in the first place. It should also be noted that any form of reciprocal dependence is entirely absent: human malaria does not support, give rise to, or constitute the mosquitos carrying it. It is a relation of asymmetrical dependence. Dependence is not a "crude" precursor of a more refined conception of causation and mechanisms in specific cases. Dependence is a macroscale, diffuse, durational, practical and theoretical conception tied directly to the reproduction and development of human and nonhuman life. Ontologically speaking, it is an independent relation that cannot be distributed into component causal parts because those parts are scattered through time and space, partially overlapping, redundant, complex, systematic, in positive and negative feedback relations.

(3) A final observation reveals the genuine significance of dependence in this context. For example, the components (parts or wholes) of ecosystem processes (e.g., water cycle) on which humans depend may be substitutable, but the processes themselves are not. This indicates that there is a structural dependence of processes upon other processes, even when the conditioning process does not directly causally determine the conditioned one. For example, social processes depend on the existence and functioning of individuals, but no single process within the individual causes social processes to occur. There is a diffuse dependence relation. Ontologically, identifying the processes and structures responsible takes place by means of categories. Therefore, we need an ontology that is capable of articulating the way that categories or principles depend on other principles. Flat or relational ontologies are not up to this task.[40]

The sets of specific categories corresponding to phenomenal strata manifest some dependence relations to each other. Each of the four domains of phenomena covered below presents a coherent but not closed network of special categories at a level of "intermediate complexity." There is a stratum of material or physical reality, including for example the categories of corporeality, space, time, process, condition, substance, causality, reciprocity, dynamic structure and equilibrium, along with other physical categories. Secondly, "vital" or organic beings embody a peculiar organic structure defined in terms of adaptability, purposiveness, metabolism or self-regulation, self-restoration, reproductive fitness, hereditary constancy and variation, genome, niche, and ecosystem, among others. Thirdly, thanks to the work of animal researchers and animal rights proponents, it is widely recognized that many types of animals possess a mental life. This "psychic" stratum includes awareness, unconscious processes, pleasure and pain, conditioned learning, habit, associative memory, communication, emotional response, problem-solving intelligence, and the categories of rigid social relations. While it is obvious that many animals are highly intelligent (such as ravens, elephants, dolphins, and nonhuman primates), it is less clear whether they have any capacity to reflect on their impulses and inclinations in an evaluative way. Capacities such as this distinguish the fourth domain. This capacity to strongly evaluate first-order motives is a characteristic difference between other-than-human animals and human being.[41] As discussed in chapter 2, this capacity is closely bound up with human language, and articulating our motivations for action in a language of values is primarily how we understand that action and ourselves to be ethical or not. We can include this capacity to strongly evaluate in the stratum of "eccentric" capacities. These include the power of conceptual thought, knowledge acquisition and creation, grasp of ideal relations, moral evaluation, symbolic communication (signification), reflectively purposive reasoning, self-aware personality, and categories bearing on the complex and variable social relationships evinced by humanity. Historical reality and culture form the immediate context for the exercise of these eccentric traits. Simply stating these categorial differences between the human and other-than-human is not to claim that humans are superior to other-than-humans. As some Panromantic pessimists and environmentalists have argued, human capacities might just as well indicate inferiority and decadence as sophistication. Whether these capacities count as good or ill depends on whether and to what extent humankind does not offend against conservative vital values and brings dynamic values into the world that would otherwise not exist without them, such as care for the Earth.

The terms *material*, *vital*, *psychic*, and *eccentric* name strata or domains of categories and concreta determined by them. The claim of the categorial ontologist is that we cannot make sense of the concreta at each respective level unless we use categories such as those listed, for they are indispensable for understanding (even if they change historically in order to better approximate the real world). The next question is what the relations among these strata of categories themselves are. Some categories recur, are modified, or are dependent on others. This requires the conception of some second-order principles that take the whole structure of the real world as their complex concretum, and these higher order categorial principles define the relations between strata. Grasping these second-order principles is the key to avoiding the errors of different varieties of environmentalist ontology and metascientific stance, and to responding appropriately to the problem of dependence denial.

The categories and strata are related in multiple ways. Most important for this book is the relation of *superposition*, which directly expresses the kind of asymmetrical (diffuse) dependence most relevant to environmentalism. The relationship of supporting and supported can be found universally throughout the real world and is the core of the relation of dependence. Here the higher categories are in most cases dependent on the lower as condition for their existence (e.g., no minds without bodies), but not for their content (e.g., bodies do not simply determine what is in the mind), and the higher level retains an independence from the lower. This relationship is most important for understanding the structure of the world, and lies at the basis of categorial ontology.[42] The fundamental categorial principle is that the lower categories on which higher strata depend are the "stronger," or invariable conditions or fundaments, while the higher are "weaker." This means that they everywhere remain the foundation and basis for the strata placed on top of them, but it does not mean that they recur in or substantively penetrate into the higher strata.[43] This fact is reflected in the idea of the *indifference* of the lower strata to that which they support. In other words, the physical world does not exist for the purpose of being taken up into the organic, just as both do not exist to "service" human beings. The physical world is indifferent to the fact that life somehow emerged from it, just as the bulk of living and nonliving natural processes could happily continue on without the existence of human beings. None of this talk of higher and lower is meant to imply the kind of evaluative hierarchical thinking that characterized ancient teleological ontology or dualism. Dependence is an ontological relation as conceived here, not an evaluative relation. None of the phenomena captured in these strata relations are more or less real than others.

This conception of categories and stratification supports my political ecological response to worldview clash and other antirealist approaches, as well as my resistance to standard universalizing science-driven environmentalism. To all forms of idealism that affirm the power of the higher categories to determine the lower, including worldview approaches, we may respond with a critique in terms of strata. If anti-realism makes eccentric categories the most fundamental, then we wind up with an epistemic anthropocentrism which devalues the real world and its structure by emphasizing the determinative role of human consciousness and expression (falling back into dualist logic). Instead of such blatant human chauvinism, we claim that the human being's "task is to come to terms with a world not made for [it]."[44] In contrast, materialist reductionism overinterprets dependence as genetic determination and attempts to explain higher phenomena in terms of lower categories alone. The conception of stratification explained here, captured in the principle of dependence, allows us to see the second-order relations between principles themselves, and it is these second-order relations that make it possible to characterize human dependence on nonhuman nature at the intermediate ecological materialist scale.

This is why stratified ontology provides a comprehensive nonanthropocentric framework for environmentalism. It helps to undermine the anthropocentric bias of a culture that thrives on its denial of dependence on nature. To sum up this contribution, I suggest that there are at least two explanatory functions that such a strata theory serves. In the context of environmentalism these functions allow us to guard against one-sided accounts of environmental problems that would place all of the responsibility for them on one type of factor (e.g., religious tradition, technological development, human chauvinism, mechanism, capitalism) despite the role of heterogeneous factors in their composition. This is the first function of such a stance. While it may be that the physical is merely a necessary existential condition for the mental and does not figure strongly into its content, it can never be entirely backgrounded and denied relevance. *Everything* has to be taken into account. We accept this point concerning ontological parity from the relationalists. More than this, according to the second global function, the laws of stratification and dependence provide a weighted order of dependence among explanatory principles themselves, which gives a macroscale structure to the ontological parity of what we started with. By means of it, we are able to recognize certain explanatory strategies as irremediably incomplete from the start.

Both can be applied in environmentalist theorizing. The most basic second-order principle is that of dependence. My references to the concept of dependence throughout the book rely on this conception. This does not mean that physical or organic relations are "more real" than all others, but that they are invariably presupposed by all "higher" principles in a strict order of dependence. Therefore, any interaction that involves human significations, for example, is invariably also a psychological event, an organic event, and a material event. "Nature" is not merely an idea, and we cannot disengage the field of signification from the supporting strata as if it were a free-floating domain without connection to the others and—having assumed in advance that humans are essentially *Homo symbolicus*—give final ontological power of determination in human experience to the play of meanings and significations.[45] How to weigh the contributions of factors in explanations without dogmatism is a crucial point for a nonreductive pluralism. That is the second global function of the theory of stratification. For the purposes of political ecology, it is important to recognize the existential dependence of higher on lower, but also the autonomy of the higher in relation to the lower. A global approach allows us to avoid the dogmatic application of one's favored "ism" while also providing an ontological framework slanted in an appropriately earthly direction. This ontology forms the nonreductive basis for critical environmental philosophy.

Human dependence on nonhuman nature is diffuse and all pervasive; it is not limited to a few parts or wholes without which we could not grow, reproduce, and thrive; it is a matter of all of these processes operating well. An example might help to show why this is the case. When we discuss an "ecosystem service" like the water cycle we know that we are referring to no particular molecule of H_2O, for many parts and wholes may be replaced or substituted over the course of many cycles. While system functions might be preserved over time, the parts and wholes that fulfill them are often replaced or changed. Human communities thus do not generally depend on any particular part-whole in their environment, but are dependent upon its self-reproducing but nonidentical processes. Only the structured process itself remains the same, but this means the principles of its operation remain the same. Indeed, it is difficult to point to some specific part of the environment and say that humans depend on just that part or whole. Human and nonhuman dependence on ecological processes is diffuse, and relates to multiple processes and systems that can seldom be readily delimited. This relation of asymmetrical dependence or superposition is a basic fact of world structure, and cannot be captured in terms of relations of container and contained between parts and wholes, nor in terms of linear serial relations between components in ongoing processes of construction and linkage. Asymmetrical dependence slips from view on these accounts, making them inadequate responses to denial of dependence. All individuals, including the human investigator, are mixed, stratified, and each becomes comprehensible only through the interrelatedness of the strata.[46]

Stratification lets us see that ontological parity of factors (interrelation or internal relations) is not enough for critical environmentalism. If we lose sight of the dependence of life on material processes, and of the mental and social on life and matter, then we have no real way of identifying priorities for practical environmentalism or political ecology. The general institutional decoupling of the economy from the Earth (chapter 5, section 3) is a prime manifestation of denial of dependence, and rectifying this denial is a crucial factor in resisting the dominant system and its mode of response to environmental problems. But resistance and intervention need to be informed by a pluralistic approach that takes into account both stratification and relation. The moral values of liberation and solidarity can be most effectively implemented when both the global order of dependence is respected and practical intervention can be directed to specific local opportunities. This order of dependence is the structure of the real world, not something imagined in a worldview, and not reducible to causal relations. Always with an eye prospectively toward the survival and reproduction of human and nonhumankind on the Earth, due account must be taken of the constraints placed on human goals by this existing order. Dependence can no longer be denied by the hegemonic system.

In the first two parts of the book, I have shown that key topics in environmental philosophy contain a host of problems needing attention, and I have responded to these problems in a manner consistent with the nondualistic principle of dependence. Stratified ontology is a necessary part of this response. Humans are situated embodied and embedded agents, dependent upon real-world structures for their continued health and flourishing, who orient ourselves in natural and social reality, epistemologically, ethically, and politically. Successful knowledge is constructed by agents from heterogeneous factors and resources, some subset of which are also epistemic values and practical goals. A critical environmental philosophy conjoins philosophical anthropology, value theory, and social-natural ontology in a political ecology that rejects conventional ways of conceiving the ecology-environmentalism relation. Stereotypical worldview clash and science-driven approaches may be avoided, and with it knowledge makers are more appropriately seen as the social actors they are. Environmentalists engage with other social actors imbricated in intersecting processes and heterogeneous webs of resources in which they co-produce knowledge of environmental-social problems, and are not passive receivers waiting to act on the dispensations of knowledge from the allegedly apolitical sciences. If human flourishing in liberated communities of solidarity with human and other Earth-others is a valuable priority, then this flourishing has to be achieved in a world that is not only structured by the direct result of our actions, but is largely not structured by humankind at all, a world not made for us. The principle of dependence encapsulates this stance.

CONCLUSION

A World Not Made for Us

THE PRINCIPLE OF DEPENDENCE—THE ASYMMETRICAL AND DIFFUSE DEPENDENCE OF HUMANKIND ON more-than-human nature—can serve as a guiding principle for reconsidering the central topics of environmental philosophy in their anthropological, epistemological, ontological, ethical, and political dimensions. The outline of a critical environmental philosophy has emerged by way of contrast with historical and existing alternatives, and I have argued that if environmentalists hope to accomplish the kinds of social change they advocate, they will have to think more deeply about the conceptual stances they take up toward environmental problems.

Among the environmental humanities disciplines, environmental philosophy is best suited for this type of work. As the discipline traditionally oriented to providing an account of "the whole," it achieves this not by exhaustive and encyclopedic empirical inventories of environmental problems, but by digging into the rich soil for the seeds—the germinal assumptions, concepts, categories, conceptual frameworks—from which sprout the many ways of characterizing the environmental problems communities deal with, as well as the frameworks for their responses to these problems. No other discipline aims at a depth survey of the whole in this way. In order to accomplish this it must be multidimensional—taking into account upstream and downstream reasoning, conventional and creative thinking, and the historical rootedness of problems. My discussion of the various aspects of the principle of dependence has probed these dimensions, and in this I hope it has been provocative for readers both new to and familiar with these ideas. I have tried to state the problems of anthropocentrism, intrinsic value, and ecological worldview in the terms provided by earlier theorists, to examine the conceptual frameworks that led to these problem statements, and to move to a larger frame of reference in each case that allows us new freedom to reconceptualize them. I found that for each problem, not only does some form of traditional dualistic thinking form one of the driving forces behind the statement of the problem itself, it continues to shape theoretical responses to it. The dualistic framework itself still has to be carefully dismantled in its various permutations in order to move toward a satisfactory resolution of these classical issues. Only environmental philosophy seems to be equipped to do this. In closing, let me briefly review the way these topics have been treated in the three major parts of the book.

If there is anything that environmental philosophers should have in common aside from the name, it is the conviction that the world we all inhabit *was not made for us*. The critique of anthropocentrism from the beginning was opposed to any conception of the human that placed it above, beyond, or outside of the nonhuman natural world in a position of dominance, control, or exploitation. Its attacks on the central chauvinist claim that the satisfaction of human interests, preferences, needs, or wants must always take priority over the interests of any other being have been persistent and trenchant. The dominion thesis, which says that the world and all it contains is "for us," was one of the first cultural myths to be debunked by environmental theorists. This problem of anthropocentrism is a manifestation of the cultural persistence of certain models of what human nature is like and of humankind's place in the world, and so immediately opens on to a much broader field: philosophical anthropology. Examining historical and current models of the human must be an essential element of any approach to rethinking human-nonhuman nature relations in environmentalism. In this case, finding existing models lacking led me to outline an ecological materialist anthropology that would be consistent with the principle of dependence. This meant endorsing a nonreductive naturalistic account of humankind as an "unfinished animal," of humans as embodied and embedded moral agents with developmental and reproductive life span who must *lead* their lives and thrive in a natural-social world not made for them. Adopting such a model reveals the shortcomings of the existing options in environmentalism, including Kantian rationalist and primitivist Panromantic variations. It avoids the extremes of human exceptionalist idealism as well as biological determinist reductivism. By emphasizing ontogeny and embeddedness—growth, reproduction, and flourishing in communities—this anthropology takes all human variation to be real variation. This does not mean that there are no cross-culturally shared aspects of our lives that stem from our embedded ecological condition. While our responses to places, living and nonliving things, need for nourishment, companionship, security, and shelter may vary, what does not vary are the structures that require us to eat, sleep, find shelter, live with others, engage in projects, produce waste, and die. This ecological materialist anthropology also provides a firm metaethical basis for environmental value theory.

It could be said that the problem of the intrinsic value of nature was also motivated by the nagging sense that the *world is not made for us*—that in contrast to "human" instrumental value there is also nonhuman nature's own intrinsic value, indifferent to us. Whether this value was subjective or objective, had its source and locus in humans or not, and could be referenced in order to put an end to disputes over priorities were central questions in the now-classical debates. In environmental philosophy the concept of the intrinsic value of nature has fallen out of favor. Here too, the inadequacy of the approaches that promoted it has to do with the conceptual framework within which it was tackled, and to get beyond it requires moving to a larger frame of reference: to value theory broadly conceived. For it, instrumental and intrinsic value are not specific value qualities at all, but placeholder terms indirectly referencing a far more differentiated field of qualities and values, including goods values, economic, vital, and moral values. The central problem with the notion of intrinsic value regarded as a considerability-conferring property is that even if some entities possess it, this fact would make little difference for the chronic experience of value conflict and clashes of values in our daily lives. For instance, although we claim to regard other humans as "ends in themselves" (possessing intrinsic value), they are victimized, exploited, oppressed, and otherwise degraded daily. In light of this experience, the problem is not to show that some part of nature or other has intrinsic value, but how to prioritize the values already implicit in the practical goals and goods aimed at in leading our lives in a world not made for us. Because human beings are embodied agents embedded in natural-social systems, how this prioritization happens in patterned ways in social contexts also requires discussion and analysis if what environmentalists hope to promote is social change that affects the ethical substance of human lives, their ethos. This means that any individualistic ethics must be supplemented by an analysis of the social institutions of consumerism and propertarianism in which ethical theories themselves are generated in many societies, and within which all humans increasingly act. I had to show that if the dominant institutional context for individual prioritizing and environmental policymaking led inevitably to commodification of nonhuman nature, and if commodification entails the contingency of what is commodified, then this establishes a contradiction at the heart of capitalist culture. It means that capitalism remains completely blind to the principle of dependence, since it regards the economy as decoupled from the real world and does not consider preservation of its processes and functions a priority. In response, political ecological ethics in community contexts preliminarily has to advocate enabling environmental practices of openness, but must also insist on the conservation of conditioning vital values without which our fullest moral lives cannot be lived, and it must also facilitate conditioned dynamic moral values such as self-realization. This ethos or pattern of priorities can be cultivated and encouraged to congeal around this synthetic principle by exemplifying it, by living differently, and by triggering "intuitions" in those around us by example as well as by rational persuasion.

Finally, if Modernist mechanism facilitates the instrumentalizing human domination of nature—implying that the world is for us—then the environmentalist demand for an alternative ecological worldview also responds to the primal sense that the *world is not made for us*. A number of environmentalists took the contrast between mechanistic and ecological worldviews to be basic to a philosophical approach to the environmental crisis, and the alternative has been presented with both idealist and materialist inflections. Within environmentalism generally there has been very little cognizance of the fact that adopting this clash of worldviews model means accepting a very specific kind of metascientific stance, and that other stances make different basic assumptions about the nature, practices, goals, and place of the sciences in society. Worldview clash generally accepts that a positivist scientific stance is not helpful for environmentalism, but does not explore the possibilities for critical reflection and social engagement in and through the scientific production of knowledge itself. This problem thus urges us once again to adopt a larger frame of reference by asking: what is the best metascientific stance for environmentalists? A stance that sets up a clash of models or unitary forces to be taken or left as a whole, or a stance that reveals the many opportunities for intervening and engaging in the ongoing processes of the production of scientific knowledge in social contexts? Opting for the latter requires exploring other forms of ontology than those currently on offer as well, since positivist materialism, process philosophies, and other forms of relational flat ontology do not do an adequate job of acknowledging the diffuse dependence of humankind on nonhuman nature. A stratified, pluralist, realist categorial ontology that recognizes distinct domains of phenomena in a structural order of dependence is put forth as a partial answer to the difficulties presented by the existing options. It recognizes the limited efficacy of worldviews in action, allows us to reject the rational-social dualism in accounts of scientific knowledge production, and enables a pluralistic account of the factors involved in any agent-in-community's engagement in these projects. Finally, it also provides a "global" framework with reference to which individuals and communities might orient their labors in an appropriate order of priority, taking care not to offend against the strongest conditioning vital values and their bearers while at the same time advancing the opportunities for all of us to live fully human lives free of oppression in communities of liberation and solidarity. It is an ontology and metascientific stance fully appropriate for a critical environmental philosophy, or political ecology in the widest sense.

Ideally, the book will provoke readers of all types to be more reflectively and practically engaged in motivating the kinds of social change that will lead to the development of an environmental culture. By beginning where we are, we might be better placed in the near future to survive the apparently inevitable collapse of an ecocidal civilization that has failed to acknowledge the principle of dependence.

NOTES

INTRODUCTION

1. A classic and still relevant text is Richard and Val Routley's "Human Chauvinism," in *Environmental Philosophy*, ed. D. S. Mannison, M. A. McRobbie, and Richard Sylvan (Dept. of Philosophy, Research School of Social Sciences, Australian National University, 1980), 96–189.
2. Warwick Fox, *Toward a Transpersonal Ecology* (Albany: State University of New York Press, 1995), 13. I have rearranged the passage for readability. At the time, Plumwood called his "the most thorough treatment of anthrocentrism in the ecophilosophical literature" ("Androcentrism and Anthrocentrism: Parallels and Politics," *Ethics and the Environment* 1, no. 2 [1996]: 132).
3. See Richard Routley, "Is There a Need for a New, an Environmental Ethic?," in *Environmental Philosophy: From Animal Rights to Radical Ecology*, 4th ed., ed. Michael E. Zimmerman, J. Baird Callicott, John Clark, Karen J. Warren, and Irene J. Klaver (Englewood Cliffs, NJ: Prentice Hall, 2004), 16–24.
4. While a periodic spurt of essays on intrinsic value occasionally appears in the literature, the spring seems to have nearly run dry. In a recent introductory text on environmental ethics written by one of the main parties involved in the debates over intrinsic value, Holmes Rolston III, the term surprisingly does not even appear in the index (*A New Environmental Ethics: The Next Millennium of Life on Earth* [London: Routledge, 2012]). Discussion of it occupies approximately 10 pages of a 225-page book. It is, however, still considered by most to be a central category of the discipline.
5. J. Baird Callicott, "Non-Anthropocentric Value Theory and Environmental Ethics," *American Philosophical Quarterly* 21, no. 4 (1984): 299.
6. Callicott, 299. See also the early statement by Tom Regan in "The Nature and Possibility of an Environmental Ethic," *Environmental Ethics* 3, no. 1 (1981): 30, 31, 34.

7   Eugene C. Hargrove, "Weak Anthropocentric Intrinsic Value," *Monist* 75, no. 2 (1992): 183.
8   Some of the sources referred to include Arne Naess, "The Shallow and the Deep, Long-Range Ecology Movement: A Summary," *Inquiry* 16, no. 1–4 (1973): 95–100, and *Ecology, Community, and Lifestyle: Outline of an Ecosophy*, chapter 2; Carolyn Merchant, *The Death of Nature: Women, Ecology, and the Scientific Revolution* (San Francisco, CA: Harper & Row, 1980); Charles Birch and John B. Cobb, *The Liberation of Life: From the Cell to the Community* (Cambridge: Cambridge University Press, 1981); J. Baird Callicott, "The Metaphysical Implications of Ecology," *Environmental Ethics* 8, no. 4 (1986): 301–16; J. Baird Callicott, "Ecology: An Ethical Perspective," *Nature Education Knowledge* 3, no. 10 (2012): 16–17; Warwick Fox, *Toward a Transpersonal Ecology: Developing New Foundations for Environmentalism* (Boston, MA: Shambhala, 1990); Bryan G. Norton, *Toward Unity among Environmentalists* (New York: Oxford University Press, 1991); Murray Bookchin, *The Philosophy of Social Ecology: Essays on Dialectical Naturalism* (Montreal: Black Rose Books, 1990); Arran Gare, *Nihilism Inc.: Environmental Destruction and the Metaphysics of Sustainability* (Como, NSW, Australia: Eco-Logical, 1996); Val Plumwood, *Feminism and the Mastery of Nature* (London: Routledge, 1993); A. Hedlund-De Witt, "Exploring Worldviews and Their Relationships to Sustainable Lifestyles: Towards a New Conceptual and Methodological Approach," *Ecological Economics* 84 (2012): 74–83; Neil Evernden, *The Natural Alien: Humankind and the Environment* (Toronto: University of Toronto Press, 1993); and David Abram, *The Spell of the Sensuous: Perception and Language in a More-Than-Human World* (New York: Vintage, 1997). For discussion of so-called Continental environmental philosophy, see Irene Klaver's introduction and notes to "Environmental Continental Philosophy" in *Environmental Philosophy: From Animal Rights to Radical Ecology*, ed. Michael E. Zimmerman et al. (Englewood Cliffs, NJ: Prentice Hall, 2004). Bruce Foltz also characterizes trends in "Environmental Continental Philosophy" in *A Dictionary of Continental Philosophy* (New Haven, CT: Yale, 2006).
9   Ian Hacking, *Representing and Intervening* (Cambridge: Cambridge University Press, 1983), xv.
10 10. Throughout, I use the terms *nonhuman*, *more-than-human*, and *other-than-human* interchangeably.
11 See especially *Environmental Culture: The Ecological Crisis of Reason* (London: Routledge, 2002), chapter 5, and *Feminism and the Mastery of Nature*, chapter 2.
12 Arne Naess, "The Deep Ecology Movement: Some Philosophical Aspects," in *Deep Ecology for the 21st Century*, ed. George Sessions (Boston: Shambhala, 1995), 64–84. This does not mean that Naess himself had such a simplistic view.

13 By the term *environmental* or *social imaginary* I mean "socially conditioned self-images, commodity images, and images of the other [, including] prevailing myths and paradigmatic narratives," following John Clark's definition in *The Impossible Community: Realizing Communitarian Anarchism* (New York: Bloomsbury, 2013), 35. Compare Lorraine Code's discussion of the category of the imaginary in her book *Ecological Thinking: The Politics of Epistemic Location* (Oxford: Oxford University Press, 2006), 29–34. For her, the imaginary "is about often-implicit but nonetheless effective systems of images, meanings, metaphors, and interlocking explanations-expectations within which people, in specific time periods and geographical-cultural climates, enact their knowledge and subjectivities and craft their self-understandings. Imaginaries are self-reinforcing rather as self-fulfilling prophesies are" (29). Both writers adopt the concept from Franco-Greek psychoanalyst and political philosopher Cornelius Castoriadis, who uses it extensively in his work. See *The Imaginary Institution of Society*, trans. K. Blamey (Cambridge: The MIT Press, 1998).

14 A critique of what comes to be called *wilderness environmentalism*, from two very different perspectives, can be found in Ramachandra Guha, "Radical American Environmentalism and Wilderness Preservation: A Third World Critique," *Environmental Ethics* 11, no. 1 (1989): 71–83, and William Cronon's "The Trouble with Wilderness; or, Getting Back to the Wrong Nature," in William Cronon, ed., *Uncommon Ground: Rethinking the Human Place in Nature* (New York: W. W. Norton, 1995), 69–90.

15 See, for instance, Paul Robbins, *Political Ecology: A Critical Introduction*, 2nd ed. (Malden, MA: Wiley-Blackwell, 2012).

16 *Anthropocentrism* is what Takis Fotopoulos calls an environmentalist myth. "According to the main myth reproduced by the system, it is 'human activity,' or 'man' in general, that are responsible for the greenhouse effect and the consequent catastrophic climatic change threatening us all." Other abstract entities have been enlisted in order to explain the crisis as well. "Another myth, which is adopted mainly by various irrational (religious and spiritualist) currents, deep ecologists, primitivists, et al., is that it is the Industrial Revolution, as well as industrial civilisation and its values, that are to be blamed for the current crisis. Similarly, others, influenced by Castoriadis's thought, blame the imaginary of development, which emerged at the same time as part of the ideology of Progress that dominated modernism in the aftermath of the Enlightenment." In the end, I agree with Fotopoulos that "both our values and our way of life are crucially determined by the prevailing socio-economic system, which is defined by the market economy and the growth economy," and "that neither a radical change in our values nor in our way of life are feasible, unless both are accompanied by a parallel change in the socio-economic institutions defining the present system." (From "Myths on the Ecological Crisis," 2007, at http://www.inclusivedemocracy.org/

journal/vol3/vol3_n02_Myths_Ecological_Crisis.htm.) However, combatting the dualism still implicit in this response has a certain priority because it shapes every attempt to explain the causes of the crisis.

17 "The Historical Roots of our Ecologic Crisis," *Science* 155, no. 3767 (1967): 1203. The recent evaluation of Anthropocene discourse by historians Christophe Bonneuil and Jean-Baptiste Fressoz, *The Shock of the Anthropocene: The Earth, History and Us*, trans. David Fernbach (London: Verso, 2017), also explicitly critiques this universalizing tendency. Additionally, historian Dipesh Chakrabarty discusses conceptual problems surrounding the use of the category *human species*, as well as a critique of universalism in the discourse of climate change, in "The Climate of History: Four Theses," *Critical Inquiry* 35, no. 2 (2009): 197–222.

18 K. S. Shrader-Frechette and Earl D. McCoy, *Method in Ecology: Strategies for Conservation* (Cambridge: Cambridge University Press, 1993); David R. Keller and Frank B. Golley, *The Philosophy of Ecology: From Science to Synthesis* (Athens: University of Georgia Press, 2000).

19 Fully developing such an anthropological view is far beyond the scope of this book. I sketch out the basic framework for such a view in regard to the principle of dependence and rejection of dualism, and display its distinctiveness relative to the many inadequate options on offer. An earlier attempt is presented in Keith R. Peterson, "All That We Are: Philosophical Anthropology and Ecophilosophy," *Cosmos and History: The Journal of Natural and Social Philosophy* 6, no. 1 (2010): 60–82.

CHAPTER ONE

1 Arran Gare, "Philosophical Anthropology, Ethics and Political Philosophy in an Age of Impending Catastrophe," *Cosmos and History: The Journal of Natural and Social Philosophy* 5, no. 2 (2009): 264.

2 Respectively, *The Enemy of Nature: The End of Capitalism or the End of the World?*, 2nd ed. (London: Zed Books, 2007), 107; *Ecological Literacy* (Albany: State University of New York Press, 1992), 18.

3 Thomas Berry, *The Great Work: Our Way into the Future* (New York: Bell Tower, 1999), 159.

4 Representative texts include Arne Naess, *Ecology, Community and Lifestyle: Outline of an Ecosophy* (Cambridge: Cambridge University Press, 1993); Murray Bookchin, *The Philosophy of Social Ecology: Essays on Dialectical Naturalism*, 2nd ed. (Montreal: Black Rose Books, 1995); Val Plumwood, *Environmental Culture: The Ecological Crisis of Reason* (London: Routledge, 2002).

5 Warwick Fox, *Toward a Transpersonal Ecology: Developing New Foundations for Environmentalism* (Boston, MA: Shambhala, 1990), 11, my emphasis. Following quote from Seed on same page.

6 Frederic Bender, *The Culture of Extinction* (New York: Humanity Books,

2003), 69. Bender regards anthropocentrism as the "fatal flaw" at the root of Christianity, Modernism, capitalism, and nihilism (19).

7   William Grey, "Anthropocentrism and Deep Ecology," *Australasian Journal of Philosophy* 71, no. 4 (1993): 469.

8   I should emphasize that contrary to popular belief, Anthropocene discourse is not really an environmentalist discourse (in the sense of attributing intrinsic value to nonhuman nature), nor is it at all critical of the conceptual framework on which it is based. It is especially guilty of the universalizing mistake identified in the introduction. By now there are numerous critical responses to the term and approach. These include Eileen Crist, "On the Poverty of Our Nomenclature," *Environmental Humanities* 3 (2013): 129–47; Dipesh Chakrabarty, "The Climate of History: Four Theses," *Critical Inquiry* 35, no. 2 (2009): 197–222; and Donna Haraway, "Anthropocene, Capitalocene, Plantationocene, Chthulucene: Making Kin," *Environmental Humanities* 6 (2015): 159–65.

9   Passmore explains very clearly how Descartes and Bacon both understood science to be the means for asserting human domination over nature. See *Man's Responsibility for Nature*, 2nd ed. (New York: Scribner, 1974), 18–22.

10 Grey, "Anthropocentrism and Deep Ecology," 463. He is apparently paraphrasing a much better original passage of Freud's from his "Introductory Lectures on Psychoanalysis." See *The Standard Edition of the Complete Psychological Works of Sigmund Freud*, vol. 16, trans. and ed. James Strachey (London: Hogarth Press, 1963), 284–85.

11 Aldo Leopold, *A Sand County Almanac: With Essays on Conservation* (New York: Oxford University Press, 1949) 204–5. In a recent book, Donna Haraway also refers to the same passage from Freud, and instead of ecology as the next stage, adds that informatics and cyborgs conjoin "organic and technological flesh" and inflict a further wound to human exceptionalism. See *When Species Meet* (Minneapolis: University of Minnesota Press, 2008), 11–12.

12 On this point, see the still unsurpassed 1979 critique written by Richard Sylvan and Val Plumwood (then both surnamed Routley), "Against the Inevitability of Human Chauvinism," in *Ethics and Problems of the 21st Century*, ed. K. E. Goodpaster and K. M. Sayre (Notre Dame, IN: University of Notre Dame Press, 1979), 36–59.

13 Paul W. Taylor, *Respect for Nature: A Theory of Environmental Ethics* (Princeton, NJ: Princeton University Press, 1986).

14 Paul W. Taylor, "The Ethics of Respect for Nature," in *Environmental Ethics: An Anthology*, ed. Andrew Light and Holmes Rolston III (Oxford: Blackwell Publishing, 2003), 82.

15 Taylor, "The Ethics of Respect for Nature," 83.

16 Bender summarizes a typical empirical argument for this position, which he calls the "natural needs" argument, in *The Culture of Extinction* (New York: Humanity Books, 2003). Empirical arguments for the inevitability of

anthropocentrism also fail for the simple historical, sociological reason that human groups have existed that have not exploited the natural world in order to survive.

17 Fox, *Transpersonal Ecology*, 20–22; Bender, *Culture of Extinction*, 82–86, meant to echo the egocentric predicament addressed by the New Realists of the early twentieth century. On these New Realists, see Roderick M. Chisholm, *Realism and the Background of Phenomenology* (Glencoe, IL: Free Press, 1960).

18 Ray Brassier, "Objects and Concepts," in *The Speculative Turn: Speculative Realisms and Materialisms*, ed. Harman et al. (Sydney: Re-Press, 2010), 47–65. The Routleys also noted how this parallels arguments for the inevitability of egoism.

19 Quentin Meillassoux, "Speculative Realism," with Ray Brassier, Iain Hamilton Grant, and Graham Harman, *Collapse* III (November 2007): 408–9.

20 "Once I distinguish the claim that my thoughts cannot exist independently of my mind, which is trivially true, from the claim that *what* my thoughts are about cannot exist independently of my mind, which simply does not follow from such a trivial truth," then I cannot be deceived by the claim that it is impossible to think of something existing independently of the mind. "The correlationist conceit is to suppose that formal conditions of 'experience' (however broadly construed) suffice to determine material conditions of reality. But that the latter cannot be uncovered independently of the former does not mean that they can be circumscribed by them" (Brassier, "Objects and Concepts," 63). The alternative to constructivism here is not a traditional naïve realism about objects, because their transobjectivity (their being more than mere objects "for us") remains open to further ontological consideration.

21 Readers may themselves compare the accounts in *Feminism and the Mastery of Nature* (London: Routledge, 1993), ch. 2; "The Politics of Reason: Towards a Feminist Logic," *Australasian Journal of Philosophy* 71, no. 4 (1993); "Androcentrism and Anthrocentrism: Parallels and Politics," *Ethics and the Environment* 1, no. 2 (1996); and *Environmental Culture*, ch. 5. Spelling in passages from Plumwood has been North-Americanized.

22 Plumwood, "Politics of Reason," 443–44. A similar list of dichotomies is presented by Jacques Derrida in *The Animal That Therefore I Am*, trans. David Wills (New York: Fordham University Press, 2009), and has been a reference point for recent work in posthuman and animal studies. The difference between them is that Plumwood reads them as power-infused, evaluative, and affectively charged oppositions, while Derrida seems to regard them in a flatter structuralist or dialectical manner.

23 Plumwood, "Politics of Reason," 443–44. List of terms from Plumwood, *Feminism and the Mastery of Nature*, chapter 2.

24 Plumwood, *Feminism and the Mastery of Nature*, 45. Many feminists theorists discuss such associations, among them Genevieve Lloyd in *The Man of Reason*

(London: Routledge, 1993) and Nancy Leys Stepan, "Race and Gender: The Role of Analogy in Science," *Isis* 77, no. 2 (1986): 261–77.
25 All of this from Plumwood, "Androcentrism and Anthrocentrism," 134–37. Internal references trimmed out.
26 Plumwood, 137–38.
27 Val Plumwood, "Nature, Self, and Gender: Feminism, Environmental Philosophy, and the Critique of Rationalism," in "Ecological Feminism," special issue, *Hypatia* 6, no. 1 (1991): 17.
28 Plumwood, *Environmental Culture*, 102.
29 Plumwood, "Androcentrism and Anthrocentrism," 135.
30 Plumwood, *Environmental Culture*, 104.
31 Plumwood, 121.
32 Plumwood, 120.
33 "Nature is represented as inessential and massively denied as the unconsidered background to technological society." Plumwood, 108.
34 Simone de Beauvoir, *The Second Sex*, trans. H. M. Parshley (New York: Alfred A. Knopf, 1971), 8.
35 Plumwood, "Androcentrism and Anthrocentrism," 138. This is related to what Weston calls "self-validating reduction" in *The Incompleat Eco-philosopher* (Albany: State University of New York Press, 2009). I'll talk more about this in chapter 4.
36 Plumwood, "Androcentrism and Anthrocentrism," 139.
37 Cf. Plumwood's critique of deep ecology in "Nature, Self, and Gender," 3–27.
38 Rosi Braidotti, *The Posthuman* (Cambridge: Polity Press, 2013), 5.
39 Braidotti, 13–15.
40 For a good account of this context, see Karl-Siegbert Rehberg's introduction to the translation of Arnold Gehlen's *Man: His Nature and Place in the World* (New York: Columbia University Press, 1988) xxxii, xxxiii–xxxiv.
41 Max Scheler, "Man and History," in *Philosophical Perspectives*, trans. O. Haac (Boston, MA: Beacon Press, 1958), 65–93.
42 Paul Taylor, *Respect for Nature: A Theory of Environmental Ethics* (Princeton, NJ: Princeton University Press, 1986), 313.
43 Paul Taylor, "The Ethics of Respect for Nature," in *Environmental Ethics: An Anthology*, ed. Andrew Light and Holmes Rolston III (Malden, MA: Blackwell, 2003), 76–77.
44 Taylor, *Respect for Nature*, 156–57. My emphasis.
45 Taylor, 49–50.
46 Taylor, 48.
47 Taylor, 121n7.
48 Taylor, 33.
49 Taylor, 36.
50 She adds that "these qualities are also identified as masculine, and hence the oppositional model of the human coincides or converges with a masculine

model, in which the characteristics attributed are those of the masculine ideal" (Plumwood, "Nature, Self, and Gender," 17).

51 See his understanding of objective and subjective in several passages (e.g., Taylor, *Respect for Nature*, 22–24). In Taylor's approach, all weight is placed on changes in individual beliefs and the power of reason, where "an *inner* change in our moral beliefs and commitments is the first, indispensable step" to social change (*Respect for Nature*, 312). He proposes no "political, legal, or economic changes in the world's present cultures" (313).

52 As Plumwood remarked, the use of the Kantian-rationalist framework "in the service of constructing a supposedly biocentric ethic" "is a matter for astonishment" ("Nature, Self, and Gender," 6).

53 Neil Evernden, *The Natural Alien: Humankind and the Environment* (Toronto: University of Toronto Press, 1993), x.

54 Evernden, *Natural Alien*, xii.

55 Evernden, 58.

56 Evernden, 59.

57 Evernden, 136. Apparently *to be* is to be *seen as*.

58 Evernden, 137.

59 Evernden, 58.

60 Plumwood, "The Concept of a Cultural Landscape: Nature, Culture and Agency in the Land," *Ethics and the Environment* 11, no. 2 (2006): 143. "In many respects the nonhuman elements rendered invisible by culture reductionism have priority as enabling, foundational conditions which make the overlay of 'cultural' elements possible" (126).

61 The "properties of consciousness ensure that whatever is perceived is not neutral, not just the passive reception of external stimuli, but the result of our active involvement in grasping the world. This makes the strictly objective world a fiction." (Evernden, *Natural Alien*, 74). More recent ecophenomenological projects accept the same basic assumptions and share many of the same weaknesses. See the collection *Eco-Phenomenology: Back to the Earth Itself*, ed. Ted Toadvine and Charles S. Brown (Albany: State University of New York Press, 2003) and Bryan Bannon, *From Mastery to Mystery: A Phenomenological Foundation for an Environmental Ethic* (Athens: Ohio University Press, 2014).

62 In other words, we have to avoid dualisms in epistemology as in other fields. Rejecting the conflation between image and entity does not entail falling back into naïve realism, since the constructivism/realism contrast obviously reflects another dualism: *either* human construct *or* natural essence. The term *transobjectivity* simply leaves open further discussion regarding the status of entities aside from what they are "for us."

63 Evernden himself perceptively acknowledges this (Evernden, *Natural Alien*, 119).

64 Other writers included under this heading would be John Zerzan, *Against Civilization: Readings and Reflections* (Port Townsend, WA: Feral House, 2005); David Abram, *The Spell of the Sensuous: Perception and Language in*

a More-Than-Human World (New York: Vintage, 1997); Kirkpatrick Sale, *After Eden: The Evolution of Human Domination* (Durham, NC: Duke University Press, 2006); and disciple of Zerzan, Derrick Jensen, whose work updates Zerzan's scathing critique of civilization without providing any productive alternative vision. See, for instance, *Endgame, Volume 1: The Problem of Civilization* (New York: Seven Stories Press, 2006) and the short film "End:Civ" partly based on it, available at https://www.youtube.com/watch?v=lvH5KFS8kfA.

65 Paul Shepard, *Coming Home to the Pleistocene* (Washington, DC: Island Press, 2004), 5–6.
66 Shepard, *Coming Home*, 1.
67 Shepard, n50, 107.
68 Shepard, 34.
69 Shepard, 77–112. The distinction between nomadism and sedentarism throughout Gilles Deleuze and Felix Guatarri's *Thousand Plateaus* (trans. Brian Massumi, Minneapolis: University of Minnesota Press, 1987) exhibits a very similar set of contrasts. These traits are listed on the table on pages 171–72.
70 For a more subtle take by an anthropologist on how this can be done, see David Graeber's *Toward an Anthropological Theory of Value: The False Coin of Our Own Dreams* (New York: Palgrave, 2001).
71 Shepard, *Coming Home*, 173. "Today we cannot become hunter-gatherers as a whole society, but we may recover some social principles, metaphysical insights, and spiritual qualities from their way of life by reconstructing it in our own milieu" (164). Even more extreme and improbable is the recent view of the bioregionalist Kirkpatrick Sale in his book *After Eden*, where he argues that in fact we should not strive to emulate early *Homo sapiens*, but should aspire to be like *Homo erectus*, who ostensibly lived in true peace and harmony with other living creatures. Sale, like Shepard, holds that all societies "'are composed of elements that are eminently dissectible, portable through time and space,' and that it is possible to 'go out or back to a culture even if its peoples have vanished, to retrieve a mosaic component, just as you can graft healthy skin to a burned spot'" (Shepard as cited in Sale, *After Eden: The Evolution of Human Domination* [Durham, NC: Duke University Press, 2006], 123).
72 Shepard, *Coming Home*, 170–73. This distinction between historical linear and mythic circular time, popular in the 1970s, is also addressed by Julia Kristeva in "Women's Time," trans. Alice Jardine and Harry Blake, *Signs* 7, no. 1 (1981): 13–35.
73 Neoteny "commits us to cultural solutions according to a calendar of development" (Shepard, *Coming Home*, 47).
74 This is one of those compromise positions scorned by anthropologist Tim Ingold. See "From Complementarity to Obviation: On Dissolving the Boundaries between Social and Biological Anthropology, Archaeology and Psychology," *Zeitschrift für Ethnologie* 123, no. 1 (1998): 21–52.

75 Shepard, *Coming Home*, 39. My emphasis.
76 Trilling quoted in Shepard, *Coming Home*, 38. My emphasis.
77 Shepard, *Coming Home*, 78. Again, "our cultural choices are rewarded or punished according to our given natures. Such constraints are part of a universal biological heritage, honed to a Pleistocene reality, to those three million years that ended about ten thousand years ago" (135; see also 146).
78 Leda Cosmides and John Tooby, "Evolutionary Psychology: A Primer," available at https://www.cep.ucsb.edu/primer.html.
79 Summary from philosopher of science John Dupré, "The Foundations of Evolutionary Psychology," in *Human Nature and the Limits of Science* (New York: Oxford University Press, 2002), 21–22.
80 John Dupré, *Humans and Other Animals* (Oxford: Clarendon, 2002), 143.
81 When we are discussing reductionism it is important to define what is meant by the term. By *reductionism* I mean, following Dupré, "the commitment to any unifactorial explanation of a range of phenomena" (*The Disorder of Things: Metaphysical Foundations of the Disunity of Science* [Cambridge: Harvard University Press, 1993], 87). Also, compare the Routleys' critique of partism in "Social Theories, Self-Management, and Environmental Problems," in *Environmental Philosophy*, ed. Don Mannison, Michael McRobbie, and Richard Routley (Canberra: Australian National University, 1980), 217–331.
82 Dupré, *Disorder*, 28. This collection of insights emerges from recent work in developmental systems theory and its allies. See, for example, Susan Oyama, Paul Griffiths, and R. D. Gray, *Cycles of Contingency: Developmental Systems and Evolution* (Cambridge, MA: The MIT Press, 2001).
83 Dupré, *Disorder*, 31.

CHAPTER TWO

1 Val Plumwood, "The Concept of a Cultural Landscape," *Ethics & the Environment* 11, no. 2 (2006): 120. Repeated in Val Plumwood, *Environmental Culture: The Ecological Crisis of Reason* (London: Routledge, 2002), chapter 2. This same dualism is addressed by Dipesh Chakrabarty in "The Climate of History: Four Theses," *Critical Inquiry* 35 (2009): 197–222. He engages in an extended discussion of philosophical anthropology in the context of climate change discourse, and struggles to reconcile historicist and naturalistic accounts of the human.
2 To be clear, I deliberately do not use the term *human nature*. What is "shared" here does not designate the nature or essence of a being, but the (fundamentally relational) conditions under which beings capable of long-term stability and change grow, reproduce, and flourish in the social-natural world. Asserting the existence of such shared conditions helps to fill in the content of the specific relation called asymmetrical dependence. It does not foreclose

further questions regarding the ultimate nature of beings, but answers to such questions have little bearing on the kinds of relations emphasized here.
3   See "Existentialism Is a Humanism," trans. Philip Mairet, in *Existentialism from Dostoyevsky to Sartre*, ed. Walter Kaufman (New York: Meridian, 1989).
4   See *The German Ideology*, in *Karl Marx: Selected Writings*, ed. Lawrence H. Simon (Indianapolis, IN: Hackett, 1994).
5   For further discussion and debate about standpoint theory, see Sandra Harding, ed., *The Feminist Standpoint Theory Reader: Intellectual and Political Controversies* (London: Routledge, 2004).
6   I use the phrases *growth, reproduction, and development* or *survival* of human life partly following Enriqué Dussel, who explains that "the '*sur*' of survival indicates, first, life from the perspective of the higher functions of the 'mind' (such as conceptual categorization, conscience, linguistic competence, self-consciousness, autonomy, etc.,) and, second, enhancement, development, new processes of innovation or cultural invention, and the creation of new conditions for human *life*" ("The Architectonic of the Ethics of Liberation: On Material Ethics and Formal Moralities," *Philosophy Social Criticism* 23, no. 1 [1997]: 24). This serves to highlight that human life is always biological-cultural life, and that there is no meaningful distinction between biological survival and its culturalized meaning for an "unfinished animal."
7   Niles Eldredge and Marjorie Grene, *Interactions: The Biological Context of Social Systems* (New York: Columbia University Press, 1992), 2–5, 68.
8   Tomasello also rejects the dualistic framework still implicitly accepted by many psychologists, where this dualistic bias results in "underappreciation of ontogeny and its formative role in the creation of mature forms of human cognition" (*The Cultural Origins of Human Cognition* [Cambridge, MA: Harvard University Press, 1999], 48). He also rejects evolutionary psychology for reasons similar to Dupre's recounted in chapter 1. "In general, the basic problem with genetically based modularity approaches—especially when they address uniquely human and socially constituted artifacts and social practices—is that they attempt to skip from the first page of the story, genetics, to the last page of the story, current human cognition, without going through any of the intervening pages" (204).
9   Tomasello, *Cultural Origins*, 11.
10 Tomasello, 79.
11 Tomasello, 79.
12 Arne Naess, *Ecology, Community, and Lifestyle: Outline of an Ecosophy* (Cambridge: Cambridge University Press, 1989), 166, 170; see also Stephen Jay Gould, "Biological Potentiality vs. Biological Determinism," in *Ever Since Darwin* (New York: W. W. Norton, 1992), 251–59. On brain plasticity as politically enabling, also see Catherine Malabou, *What Should We Do with Our Brain?*, trans. S. Rand (New York: Fordham University Press, 2008).
13 For readers already familiar with Gehlen, a note on Gehlen's politics might

be in order: Karl-Siegbert Rehberg rightly notes that, while many elements of Gehlen's thought were formed out of a politically conservative perspective, there is nothing intrinsically conservative about the anthropological framework that he develops. He says that based on Gehlen's own findings "one could also identify possibilities for the emancipation of the human being. This would of course not signify a liberation from the anthropological facts, but would instead focus on the versatility, the openness, and 'nondetermined' nature of the human species" (in the introduction to Arnold Gehlen's *Man: His Nature and Place in the World*, trans. Clare McMillan and Karl Pillemer [New York: Columbia University Press, 1988], xxxvi). Axel Honneth and Hans Joas also recognize the tenuous connection between Gehlen's anthropological ideas and his implicit political proposals, and read him against the grain of the latter, as do I. See their *Social Action and Human Nature*, trans. Raymond Meyer (Cambridge: Cambridge University Press, 1988), 48–70.

14 Gehlen, *Man*, 4.
15 Gehlen, 109. This view is directly opposed to the hypothesis of evolutionary psychologists that there is a uniform "environment of evolutionary adaptation" for the human species. The concept of the environment of evolutionary adaptation is a hypothetical model that refers to common problems that our ancestors had to solve independent of geographical location, not an evolutionary reality. According to Cosmides, "It is the statistical composite of selection pressures that caused the design of an adaptation" ("A Primer on Evolutionary Psychology," last modified January 13, 1997, http://www.cep.ucsb.edu/primer.html).
16 Eldredge and Grene, *Interactions*, 179.
17 He refers to the work of Dutch anatomist Lodewijk Bolk (1866–1930). For a good discussion situating and explaining Bolk's work, see Gould's *Ontogeny and Phylogeny* (Cambridge, MA: Harvard University Press, 1977), 352–65. On retarded growth and neoteny in general, see 365–404.
18 Gehlen, *Man*, 101.
19 Gould, *Ontogeny and Phylogeny*, 365, 399.
20 Gould, 399.
21 Gould, "Biological Potentiality vs. Biological Determinism," 464.
22 Gehlen, *Man*, 101–2. Bruce Wexler would concur with Gould, Grene, and Gehlen about the significance of neuroplasticity, and the importance of cultural scaffolding: "I suggest that the phylogenetic emergence of human beings rests, to a significant degree, on selection for an extended period of postnatal plasticity in the fine-grained shaping of the structural and functional organization of the human brain.... The extended period of postnatal neuroplasticity is an aspect of human nature that allows and requires environmental input for normal development. Moreover, since human beings provide an important part of the environmental input, one person's nature is another person's nurture" (*Brain and Culture: Neurobiology, Ideology, and Social Change* [Cambridge, MA:

The MIT Press, 2006], 16). Brain plasticity is extensively discussed in related and illuminating ways by Catherine Malabou in *What Should We Do with Our Brain?*, trans. Sebastian Rand (New York: Fordham University Press, 2008).

23 See Grene's account of Portmann's thinking in her *Approaches to a Philosophical Biology* (New York: Basic Books, 1968), as well as Eldredge and Grene, *Interactions*, chapter 7. Gould briefly discusses Adolf Portmann in *Ontogeny and Phylogeny*, 369–70.

24 Eldredge and Grene, *Interactions*, 183.

25 Gehlen, *Man*, 109.

26 Gehlen, 16.

27 My translation from *Urmensch und Spätkultur: Philosophische Ergebnisse und Aussagen* (Bonn: Athenäum Verlag, 1956), 8. In *Man* he also characterized this as a deliberately psychophysically neutral approach. The concept of action does not "raise insoluble metaphysical problems such as that of the body and soul.... The categories we will employ, of relief, communication, retardation (juvenilization), and so forth, may be 'psychophysically neutral,' as Scheler says.... This has the advantage of enabling us to avoid the body-soul problem as long as we confine our examination strictly to the phenomena themselves" (109).

28 Quoted in Gould, *Ontogeny and Phylogeny*, 401.

29 Marjorie Grene, *A Philosophical Testament* (Chicago, IL: Open Court, 1995).

30 For a more recent account of Gibson's work and its potential for environmental philosophy, see William M. Mace, "James J. Gibson's Ecological Approach: Perceiving What Exists," *Ethics & the Environment* 10, no. 2 (2005): 195–216. For an intriguing use of Gibson's work that attempts to establish a basis for intrinsic value in environments, see Mark Rowlands, *The Environmental Crisis: Understanding the Value of Nature* (New York: St. Martin's, 2000).

31 Maurizio Ferraris, *Introduction to New Realism*, trans. Sarah de Sanctis (London: Bloomsbury, 2015), 46.

32 Naess, *Ecology, Community and Lifestyle*, 67. *Designation* is contrasted with *declarative sentence*, which probably has some bearing on the truth value. See also "The World of Concrete Contents," *Inquiry: An Interdisciplinary Journal of Philosophy and the Social Sciences* 28 (1985): 417–28, where the arguments were first laid out.

33 Nicolai Hartmann, *Aesthetics*, trans. Eugene Kelly (Berlin: De Gruyter, 2014), 50. He continues, "For that reason we must say that they are given to us in the form of properties of the object, and not in the form of subjective additions (which, considered in themselves, they may largely be); not as elements of acts, but rather entirely as elements of the content of objects." Detailed explanation of this view of value properties cannot be provided here, but will be taken as a viable option and built on in later chapters.

34 Hartmann continues: "The relatedness of things to us, which is rooted in our dependence upon them, is hence in no wise an illusion, but concrete reality" (*Aesthetics*, 50).

35 Wexler, *Brain and Culture*, 5.
36 Ecologist William Rees has put this insight to use in explaining the obstacles to generating progressive climate change policies. See "What's Blocking Sustainability? Human Nature, Cognition, and Denial," *Sustainability: Science, Practice, & Policy* 6, no. 2 (2010): 13–25; and "Human Nature, Eco-Footprints, and Environmental Injustice," *Local Environment* 13, no. 8 (2008): 685–701.
37 "Intentional agents are animate beings who have goals and who make active choices among behavioral means for attaining these goals, including active choices about what to pay attention to in pursuing these goals" (Tomasello, *Cultural Origins*, 68). He distinguishes living, "animate beings" from intentional agents. The former can "make things happen," but do not necessarily show evidence of goal-oriented, anticipatory action.
38 Tomasello, 97.
39 Tomasello, 84ff. Ferraris also uses Gibson's notion of an "affordance" in a very broad sense to refer to the information that objects in the world give to perceivers—the world affects thought and is prestructured. See Ferraris, *Introduction to New Realism*.
40 Tomasello, *Cultural Origins*, 84–85. This can apply just as well to children observing adults dealing with nonhumans in particular ways—with care and respect, or with disdain and cruelty.
41 Eldredge and Grene, *Interactions*, 182–83.
42 Gehlen, *Man*, 108.
43 Tomasello, *Cultural Origins*, 212.
44 Paul W. Taylor, *Respect for Nature: A Theory of Environmental Ethics* (Princeton, NJ: Princeton University Press, 1986), 9.
45 Gehlen, *Man*, 49. Emphasis in original.
46 "Human impulses should thus be considered from the point of view of their role in the [changing] context of action" (Gehlen, 47).
47 Gehlen, 324. Incidentally, Gehlen's critique of Freud's drive theory was that it was far too narrow. Freud was right that there is overstimulation and all kinds of impulses, but he was wrong to consider them all originally libidinal or sexual.
48 Gehlen argues that it is wrong to begin by reading backwards off real behavior to deterministic predispositions for action, and to contrast with this store of ready-made impulses a given environment in which they play themselves out (321–22).
49 Gehlen, 328.
50 As Gehlen remarked, "It is inconceivable that a clear boundary could exist in [human being] between those actions that serve immediate biological purposes and those that serve indirect, removed purposes" (Gehlen, 330). "We may argue further that there is no objective boundary between impulses and

habits, between primary and secondary needs; instead, whenever such a distinction does appear, it is made by *man himself*" (Gehlen, 330).
51 Gehlen, 330.
52 Herbert Marcuse explored the idea of "false needs" in great detail in *One Dimensional Man: Studies in the Ideology of Advanced Industrial Society* (London: Routledge, 2008 [originally published 1968]), as do other Frankfurt school theorists.
53 Gehlen, *Man,* 30–31. It follows that Gehlen rejects an "objective list" approach to needs. See 323–24, 326, 328, 342.
54 Wexler, *Brain and Culture,* 47.
55 More will be made of this point in a later chapter in relation to value theory and framed by Maurizio Ferraris's conception of inscription in his social theory of documentality, developed in *Documentality: Why It Is Necessary to Leave Traces,* trans. Richard Davies (New York: Fordham University Press, 2013).
56 The indeterminateness of human motives or values is initially managed by the frontal lobes of our parents and caregivers during the period when our brains are still developing. Since children do not have fully developed centers for planning, attention, organization, and strategy selection, Wexler claims that "when the child's frontal lobes are developing, the parents' brains provide frontal lobe functions for the child" (*Brain and Culture,* 109). The processes of instrumental parenting, imitation, identification, and play below are also discussed in Wexler, *Brain and Culture,* 98–137.
57 Gehlen, *Man,* 48.
58 Tomasello, *Cultural Origins,* 108.
59 Tomasello, 118. "In all cases, then, the use of a particular linguistic symbol implies the choice of a particular level of granularity in categorization, a particular perspective or point of view on the entity or event, and in many cases a function in context" (119).
60 Tomasello, 9. Compare Gehlen: "The things themselves take on *different but equally possible values* depending on the different purposes for which or the patterns of movement in which they are used. For example, depending on the situation and intention, a stick may be used for pointing, for support, or for striking; and it will attract our attention from any of these perspectives which we may adopt toward it" (*Man,* 213).
61 Tomasello, 9.
62 "Linguistic symbols thus free human cognition from the immediate perceptual situation" (Tomasello, *Cultural Origins,* 9).
63 Tomasello, 9.
64 Tomasello, 9.
65 "Man's [sic] impulses represent a field of conflict, because different groups of heterogeneous impulses are always competing with each other to find expression in behavior" (Gehlen, *Man,* 390).
66 Another writer who has thought much about the biocultural nature of human

motivations, as well as environmentalism, is Mary Midgley. She also believes that our condition is one of a constant clash of impulses. "No motive is an infallible moral imperative, not even the persistent ones which have been most often treated in this way, such as tribal loyalty or family affection. The mere fact that a motive occurs persistently, in our own or any other species, does not give it automatic authority or turn it into a moral rule. *What makes rules necessary is the fact that motives clash, and clash in the context of a mental life that badly needs to work as a whole.* Having, apparently, more memory, foresight and imagination than other earthly creatures, we are aware, however dimly, of our lives as wholes, and of the way in which serious conflicts disrupt those wholes" (*The Ethical Primate: Humans, Freedom, and Morality* [New York: Routledge, 1996], 138). She discusses clashes and priorities throughout the book, e.g., 138–39, 144, 149, 161, 178, 181.

67 For a concise treatment of "articulation" and moral values, see "What is Human Agency?" in *Human Agency and Language: Philosophical Papers*, vol. 1 (Cambridge: Cambridge University Press, 1985), 15–43. I accept the cultural-realist interpretation of Taylor's work as it is understood by Finnish philosopher Arto Laitinen, as explained in *Strong Evaluation without Moral Sources: On Charles Taylor's Philosophical Anthropology and Ethics* (Berlin: Walter de Gruyter, 2008).

68 Charles Taylor, "Language and Human Nature," in *Human Agency and Language: Philosophical Papers*, vol. 1 (Cambridge: Cambridge University Press, 1985), 232–33.

69 Taylor, "What is Human Agency?," 25.

70 Charles Taylor, "Theories of Meaning," in *Human Agency and Language: Philosophical Papers*, vol. 1 (Cambridge: Cambridge University Press, 1985), 270–73.

71 Scheler states that "anything of positive value ought to be, and anything of negative value ought not to be.... Every ought has its foundation in values." *Formalism in Ethics and Non-formal Ethics of Values* (Evanston, IL: Northwestern University Press, 1973), 206. I simply accept Scheler's starting point here without further defending it.

CHAPTER THREE

1 As Max Scheler noted, the doctrine of value subjectivism in the moral realm corresponds with a strictly human-centered labor theory of value in political economy; in turn, these correlate with the idealist-constructivist tendency in epistemology. *Ressentiment*, trans. Lewis B. Coser (Milwaukee, WI: Marquette University Press, 1994), 99, 102.

2 John O'Neill, "The Varieties of Intrinsic Value," *Monist* 75, no. 2 (1992): 119–37. "To hold an environmental ethic is to hold that non-human beings and states of affairs in the natural world have intrinsic value" (119).

3 O'Neill, 119–20. See also his "Meta-Ethics," in *Companion to Environmental*

Philosophy, ed. Dale Jamieson (Malden, MA: Blackwell, 2001), 163–76. See also John O'Neill, Alan Holland, and Andrew Light, *Environmental Values* (New York: Routledge, 2010), 114–16.

4   Weston also astutely noted that "even the best nonanthropocentric theories in contemporary environmental ethics are still profoundly shaped by and indebted to the anthropocentrism that they officially oppose" (Anthony Weston, *The Incompleat Eco-philosopher: Essays from the Edges of Environmental Ethics* [Albany: State University of New York Press, 2009], 23).

5   J. Baird Callicott, "Intrinsic Value, Quantum Theory, and Environmental Ethics," *Environmental Ethics* 7, no. 3 (1985): 161–62.

6   Callicott, 162.

7   Holmes Rolston, "Are Values in Nature Subjective or Objective?," in *Philosophy Gone Wild: Essays in Environmental Ethics* (Blue Ridge Summit, PA: Prometheus Books, 1986), 107.

8   Holmes Rolston, "Value in Nature and the Nature of Value," in *Philosophy and the Natural Environment*, ed. Robin Attfield and Andrew Belsey, Royal Institute of Philosophy Supplement 36 (Cambridge: Cambridge University Press, 1994), 25.

9   Rolston, "Are Values in Nature Subjective or Objective?," 114. He appeals to a naturalistic anthropology which sees that "all the organs and feeling mediating value—body, senses, hands, brain, will, emotion—are natural products. Nature has thrown forward the subjective experiencer quite as much as the world that is objectively experienced" (101).

10 Rolston, "Value in Nature and the Nature of Value," 15.

11 Further, "a living thing can be said to flourish if it develops those characteristics which are normal to the species to which it belongs in the normal conditions for that species" (O'Neill, "Varieties," 129).

12 Keekok Lee, "The Source and Locus of Intrinsic Value: A Reexamination," *Environmental Ethics* 18, no. 3 (1996): 297–309; page numbers here refer to reprint in Holmes Rolston III and Andrew Light, eds., *Environmental Ethics: An Anthology* (Malden, MA: Blackwell, 2003), 155.

13 Anthony Weston, "Beyond Intrinsic Value: Pragmatism in Environmental Ethics," *Environmental Ethics* 7, no. 4 (1985): 321–39.

14 Bryan G. Norton, "Epistemology and Environmental Values," *Monist* 75, no. 2 (1992): 215.

15 Bryan G. Norton, *Toward Unity among Environmentalists* (New York: Oxford University Press, 1991), 235.

16 Eugene Hargrove, "Weak Anthropocentric Intrinsic Value," *Monist* 75, no. 2 (1992): 236.

17 Hargrove, 196. The conception of "attributed inherent value" remains equivocal in the sense that it does not tell us whether its user believes such value has its root in individual subjects (Callicott) or whether it is rooted in the collective culture (pragmatists).

18 Keekok Lee also refers to a frequent confusion of these two different questions. See "The Source and Locus of Intrinsic Value: A Reexamination," in *Environmental Ethics: An Anthology*, ed. Holmes Rolston III and Andrew Light (Malden, MA: Blackwell, 2003), 161.

19 Kenneth Goodpaster, "On Being Morally Considerable," *Journal of Philosophy* 75, no. 6 (1978): 308–25; reprinted in *Environmental Philosophy: From Animal Rights to Radical Ecology*, ed. Michael Zimmerman et al. (Upper Saddle River, NJ: Prentice Hall, 2005), 53–66. Citations refer to these page numbers.

20 Goodpaster, "On Being Morally Considerable," 55. Emphasis in original.

21 Goodpaster, 65.

22 Goodpaster, 56.

23 Cf. Tom Regan in "Animal Rights, Human Wrongs," *Environmental Ethics* 2 (1980): 104, and "Does Environmental Ethics Rest on a Mistake?," *Monist* 75, no. 2 (1992): 167, 175; Mary Midgley in "Duties Concerning Islands," *Environmental Ethics*, ed. Robert Elliot (Oxford: Oxford University Press, 1995), 97; and Christopher Stone in "Moral Pluralism and the Course of Environmental Ethics," in *Environmental Ethics: An Anthology*, ed. Holmes Rolston III and Andrew Light (Malden, MA: Blackwell, 2003), 194.

24 Incidentally, Continental approaches like Levinas's moral phenomenology of the "face" of the Other are also propertarian. Applying Levinasian concepts to environmental issues has led directly to the obvious question whether animals have "faces" in the Levinasian sense, which leads to a cul-de-sac no less pernicious and distracting than the discussions of intrinsic value. The principle is identical to that of standard moral extensionism: "widening of the constituency of the other to whom I owe responsibility." See John Llewelyn, *The Rigor of a Certain Inhumanity: Toward a Wider Suffrage* (Bloomington: Indiana University Press, 2012) 1, 288. Extend the circle to the size of the cosmos, to Tom Birch's "universal consideration," and we are still in no position to understand how to prioritize the claims of one being over another in any reproducible normative way.

25 Singer noted this specific conflict, but cast it in utilitarian terms, thereby submitting these qualitatively different values to the single metric of "utility" which, like "instrumental value," is not a specific value quality at all in my view. "All Animals are Equal," in Peter Singer, *Applied Ethics* (Oxford: Oxford University Press, 1986), 222.

26 Anthony Weston, "Between Means and Ends," *Monist* 75, no. 2 (1992): 244.

27 Val Plumwood, *Feminism and the Mastery of Nature* (London: Routledge, 1993), 186. My emphasis.

28 Weston, *Incompleat Eco-philosopher*, 35–36. Emphasis in original.

29 Weston, 66–67. Birch's idea of "universal consideration" entails resisting the biases of stereotyping and remaining open to communicative interaction with everything. The burden of proof is on those who would restrict this interaction, rather than on those who would allow it to remain open (Weston, 70).

30 Weston, 35.
31 Weston, 29.
32 Weston, 33–34.
33 Ecofeminist Lori Gruen makes a similar point, and advocates a pluralistic theory of values that "assesses" them in a public context. "Values and the reasons that capture them are molded and shaped in communities.... If the process of valuing is to some degree public, then the process of justifying values and identifying reasons should seek to be as democratic, inclusive, and informed as possible" ("Refocusing Environmental Ethics: From Intrinsic Value to Endorsable Valuations," *Philosophy & Geography* 5, no. 2 (2002): 161–62.
34 Weston, *Incompleat Eco-philosopher*, 80.
35 O'Neill, Holland, and Light, *Environmental Values*, 109–10.
36 Cheney, "Eco-Feminism and Deep Ecology," *Environmental Ethics* 9, no. 2 (1987): 142. On the place of narrative in feminist ethics, see also Karen Warren's "The Power and Promise of Ecofeminism Revisited," in *Environmental Philosophy: From Animal Rights to Radical Ecology*, ed. Zimmerman et al. (Upper Saddle River, NJ: Prentice Hall, 2004), 252–79.
37 She lists various kinds of ethics appropriate for some contexts but not for others, including "virtue ethics, care ethics, solidarity and friendship ethics, ecological and food web ethics of reciprocity, and communicative ethics" (Val Plumwood, *Environmental Culture: The Ecological Crisis of Reason* [London: Routledge, 2002], 188).
38 Plumwood, 188.
39 Plumwood, 177.
40 See René Descartes, "From the Letters of 1646 and 1649," in *The Philosophical Writings of Descartes*, ed. John Cottingham, Anthony Kenny, Dugald Murdoch, Robert Stoothoff (Cambridge: Cambridge University Press, 1991).
41 Plumwood, *Environmental Culture*, 191. This point already makes it possible for us to sharply distinguish "dialogical communicative ethics" from a project like Habermasian "discourse ethics," which builds rationalist anthropocentrism into its a priori structures of communication in a perfectly Kantian way. (See Jürgen Habermas, "Discourse Ethics: Notes on a Program of Philosophical Justification," in *Moral Consciousness and Communicative Action*, trans. Christian Lenhardt and Shierry Weber Nicholsen [Cambridge, MA: The MIT Press, 1990], as well as "Remarks on Discourse Ethics," in *Justification and Application*, trans. Ciaran Cronin [Cambridge, MA: The MIT Press, 1993], especially 110–11, where human duties to animals are only "analogous" to moral duties to humans, even if animals are conceived as communicative agents.)
42 Plumwood, *Environmental Culture*, 189.
43 The concept of "communicative action" here combines a number of threads. (1) The communicative-embodied circular processes of learning discussed in

part 1, which may or may not be linguistic; communicative action includes mute practices and conduct that gives evidence of prioritization; (2) communicative linguistic expression, such as narrative and articulation of value (Habermasian communicative action falls under this heading); (3) communicative actions (such as advertising, news media, or state policies) that obstruct or close off other kinds of communicative possibilities; (4) communicative or dialogical ethics that is open to Other faces, voices, and agencies.

44 Plumwood, *Environmental Culture*, 192.
45 Plumwood, 193.
46 She includes it among a host of others, such as "de-homogenisation of both 'nature' and 'human' categories," "minimising interspecies ranking and ranking contexts," "'studying up' in problem contexts (self-critical stance)" (Plumwood, *Environmental Culture*, 194–95).
47 Karen Warren, "The Power and Promise of Ecological Feminism," *Environmental Ethics* 12 (1990): 126–46; reprinted in *Environmental Ethics: What Really Matters, What Really Works*, ed. David Schmidtz and Elizabeth Willott (New York: Oxford University Press, 2002) 234–37; citation on 239–40. References to these page numbers.
48 Val Plumwood, "Nature, Self, and Gender: Feminism, Environmental Philosophy, and the Critique of Rationalism," in "Ecological Feminism," special issue, *Hypatia* 6, no. 1 (1991): 7.
49 Plumwood, *Feminism and the Mastery of Nature*, 183–88.
50 Plumwood, 185. Rolston makes the same point in "Environmental Virtue Ethics: Half the Truth but Dangerous as a Whole" in *Environmental Virtue Ethics*, ed. Ronald Sandler and Philip Cafaro (Lanham, MD: Rowman and Littlefield, 2005), 61–78.
51 Plumwood, 185–86. In a recent book, Lori Gruen has developed the concept of entangled empathy as a key virtue for animal ethics. See *Entangled Empathy: An Alternative Ethic for our Relationships with Animals* (New York: Lantern Books, 2015).
52 Brian Treanor has also done important work developing a sophisticated environmental virtue ethics in *Emplotting Virtue: A Narrative Approach to Environmental Virtue Ethics* (Albany: State University of New York Press, 2014).
53 For an introduction to and commentary on this tradition, see the wonderfully brief text by J. N. Findlay, *Axiological Ethics* (London: Macmillan, 1970). One might object at the outset that none of the writers listed above are environmentalists, and most of them may be human exceptionalists or idealists, so using their work in the service of environmental ethics may seem contradictory. It is certainly true that on this account we have to avoid whatever anthropocentrism, idealism, intuitionism, nonnaturalism, and intellectualism these accounts espouse. With all of the preparatory work of the previous chapters in place I think this can be done, and we will be able to add important

aspects of the axiological approach to our toolbox as an essential piece of a pluralistic value theory.

54 This list of features follows Eva Cadwallader, a student of Wilhelm Werkmeister, himself a student of Nicolai Hartmann. See "The Main Features of Value Experience," *Journal of Value Inquiry* 14, no. 3–4 (1980): 229–44.

55 Nicolai Hartmann, *Moral Phenomena: Vol. 1 of Ethics*, trans. Stanton Coit (New Brunswick, NJ: Transaction Publishers, 2002), 224.

56 Scheler, like Rolston, widened the class of valuers to encompass all conative beings, including all living things. Any being that exhibits goal-oriented behavior is considered a valuer, though not for this reason a moral valuer or moral agent. See *The Human Place in the Cosmos*, trans. Manfred S. Frings (Evanston, IL: Northwestern University Press, 2009). Hans Joas also discusses this point in *The Genesis of Values*, trans. Gregory Moore (Chicago, IL: University of Chicago Press, 2000), 91–92. For more on Scheler's view in relation to environmental values, see my "Bringing Values Down to Earth: Max Scheler and Environmental Philosophy," *Appraisal: The Journal of the Society for Post-Critical and Personalist Studies, Re-Appraisal: Max Scheler (Pt 2)* 8, no. 4 (2011): 3–12.

CHAPTER FOUR

1 Readers of Argentine philosopher Enriqué Dussel's work might recognize a resemblance between his "universal material ethical principle" and this survival principle. I have drawn much inspiration from Dussel's work, even if I disagree with him in detail. His staggeringly erudite and masterful treatment of a wide range of issues in ethical theory in *Ethics of Liberation in the Age of Globalization and Exclusion* (trans. Alejandro A. Vallega et al. [Durham, NC: Duke University Press, 2013]) is a must-read for all ethical theorists and political philosophers.

2 Petroculture has received increasing attention from humanists lately. A good introduction to and collection of some of this work is Imre Szeman and Dominic Boyer's *Energy Humanities: An Anthology* (Baltimore, MD: Johns Hopkins University Press, 2017); a provocative philosophical treatment of the multifaceted nature of human dependence on oil is provided by Finnish writers Antti Salminen and Tere Vadén in *Energy and Experience: An Essay in Nafthology* (Chicago, IL: MCM', 2015).

3 This layered relation between goods and moral values was called "Scheler's Law" by his colleague in Köln, Nicolai Hartmann (*Aesthetics*, trans. Eugene Kelly [Berlin: De Gruyter, 2014], 338).

4 See previous chapter's section "On Value Theory." Much of the discussion in this section, and of value theory in general, is indebted to Hartmann. I draw from his book *Ethics* (trans. Stanton Coit, 3 vols. [New Brunswick, N.J: Transaction Publishers, 2002–2004]) as well as his late *Aesthetics* throughout.

On the distinction and relations between goods and moral values specifically, see *Aesthetics*, trans. Eugene Kelly (Berlin: De Gruyter, 2014), 331–38.

5   Hartmann, *Aesthetics*, 331. Cf. John O'Neill, "The Varieties of Intrinsic Value," *Monist* 75, no. 2 (1992), on this point.

6   Hartmann, 332. Goods values can also depend on other values or concrete qualities. The sugar in the gas tank of the bulldozer is used because of its particulate quality of clogging, not because of its chemical properties (sugar does not dissolve in gasoline). Qualities of the object also affect the degree of moral merit or blame received for an action. Environmentalists who preserve the pangolin (scaly anteater native to Southeast Asia) may be seen as more merit worthy to some than those who protect tigers, given the less charismatic appeal of pangolins to most humans.

7   I distinguish between "goal-directedness"—which many living things possess—and "purposive behavior," or value orientation, where the value of goals may be recognized-articulated. In every activity some specific good is aimed at, whether simply mutely enacted or also recognized-articulated. Deliberately prioritizing among ends is one level of reflection above simply acting on first-order desires, but does not necessarily include the highest order of reflection that asks "is it right that I select this valued goal and prioritize it this way?"

8   The distinction between "recognized-articulated" and "mutely enacted" value comes from Keekok Lee. See "The Source and Locus of Intrinsic Value: A Reexamination," *Environmental Ethics* 18, no. 3 (1996): 297–309.

9   Thomas H. Birch, "Moral Considerability and Universal Consideration," *Environmental Ethics* 15, no. 4 (1993): 313–32.

10 This revision of value theory is inspired by Maurizio Ferraris's discussion of "the exemplarity of the sample" in *Documentality: Why It Is Necessary to Leave Traces*, trans. Richard Davies (New York: Fordham University Press, 2013), 12–13, 48–49.

11 Compare Rolston's discussion in his recent text, *A New Environmental Ethics: The Next Millennium of Life on Earth* (London: Routledge, 2012), 115–16.

12 Hartmann uses the familiar example of theft to illustrate: "In what way does the action of the honest man differ from that of the thief with respect to the unprotected property of others? In this way: the first man respects the possessions of others, while the second man does not. But the presupposition here is that the objects possessed by another have a value, specifically a goods-value, for which one may desire them.... The moral values, and likewise the moral disvalues, are thus *conditioned by* the goods-value, that is, they are founded upon it" (*Aesthetics*, 337). See also 336–38.

13 Compare Maurizio Ferraris's claim in his *Introduction to New Realism*, "Without the positivity of objects no morality is possible" (trans. Sarah de Sanctis, [London: Bloomsbury, 2015], 49).

14 Hartmann, *Aesthetics*, 337.

15 *Unless* the context would specially motivate the agent to conform to what other immoral agents are doing, and the agent must *resist* the inclination to violate it in the way that others are violating the norms, giving rise to courage or fortitude as a merit.

16 Hartmann, *Ethics*, vol. 2, 451–52, and on the other points in this paragraph, 450–51.

17 For an overview and discussion, see Baird Callicott's papers "Animal Liberation: A Triangular Affair" and "Animal Liberation and Environmental Ethics: Back Together Again" (reprinted in *In Defense of the Land Ethic* [Albany, NY: State University of New York Press, 1989], 15–38 and 49–60); and Gary Varner's "Can Animal Rights Activists be Environmentalists?," in *Environmental Philosophy and Environmental Activism*, ed. Don Marietta and Lester Embree (Lanham, MD: Rowman & Littlefield, 1995), 169–201.

18 See Callicott's "Holistic Environmental Ethics and the Problem of Ecofascism," in *Environmental Philosophy: From Animal Rights to Radical Ecology*, 4th ed., ed. Michael Zimmerman et al. (Upper Saddle River, NJ: Pearson Prentice Hall, 2005), 116–29; Wayne Ouderkirk and Jim Hill, eds., *Land, Value, and Community: Callicott and Environmental Philosophy* (Albany, NY: State University of New York Press, 2002), 298–99; and Wenz's *Environmental Justice* (Albany, NY: State University of New York Press, 1988), chapter 14.

19 Hartmann, *Ethics*, vol. 2, 94.

20 All goods values are in a sense "conditioning" from the standpoint of morality. We can certainly not be concerned at all with evaluating the intentions of others, but as soon as we adopt this second-order perspective, the goods values they aim at in purposive behavior are recognized as conditioning—necessary for having a moral life at all, but not constituting its full moral content.

21 Kant argued that "to use life for its own destruction, to use life for producing lifelessness, is self-contradictory" (Kant's *Lectures on Ethics*, cited in Enriqué Dussel, "Principles, Mediations, and the 'Good' as Synthesis," *Philosophy Today* 41 [Supplement 1997]: 59). This means that suicide is logically and *performatively* self-contradictory. "Suicide is performatively contradictory … because it extinguishes subjectivity itself or the existence of ethical beings insofar as they are real."

22 Hartmann, *Ethics*, vol. 2, 402. This is because height pertains to intentions on the backs of which moral values arise, the objects (goods) of which may remain unactualized. This is why nonactualization of a goods value (e.g., the rotten apple in the example earlier) does not affect the height of moral value per se or serve as a criterion of moral value.

23 Hartmann, 455.

24 Hartmann, 420.

25 "Not to violate the lower values and at the same time to actualize the higher" (Hartmann, 456, 458). The ideal is described in Hartmann 460–63.

26. Vandana Shiva, "The Impoverishment of the Environment: Women and Children Last," in *Ecofeminism* (Atlantic Highlands, NJ: Zed Books, 1993), 70–90.
27. Anthony Weston, "Multicentrism: A Manifesto," *Environmental Ethics* 26, no. 1 (2004): 25–40.

CHAPTER FIVE

1. Michael J. Watts, "Political Ecology," in *A Companion to Economic Geography*, ed. E. Sheppard and T. Barnes (Oxford: Blackwell, 2000), 257. For an extensive discussion of various definitions of political ecology and the problems it typically tackles see Paul Robbins, *Political Ecology*, 2nd ed. (Oxford: Wiley-Blackwell, 2012), 14–24.
2. John P. Clark, personal communication.
3. Murray Bookchin, "What is Social Ecology?," in *Environmental Philosophy: From Animal Rights to Radical Ecology*, ed. Michael E. Zimmerman et al. (Upper Saddle River, NJ: Prentice Hall, 2004), 472. This is a historical and causal claim for Bookchin.
4. John P. Clark, *The Impossible Community: Realizing Communitarian Anarchism* (New York: Bloomsbury, 2013).
5. *Environmental Philosophy: From Animal Rights to Radical Ecology*, 4th ed. (Upper Saddle River, NJ: Prentice Hall, 2004), 361–495.
6. Clark, *Impossible Community*, 23.
7. Clark, 34.
8. Clark, 36. Because the "quasi-hegemonic ideological sector in the present era is that of economistic ideology," aspects of reality are constantly cast in terms of economic metaphors and analogies. Ecological processes can easily be regarded as commodified "ecosystem services" with a price tag, as I explain in section 3.
9. Clark, 35.
10. Clark, 35, my emphasis.
11. Clark, 34–35.
12. Clark, 71.
13. He continues: "It points to the conclusion that an effective movement for social transformation must consist of a growing community whose members are in the process of creating for themselves a different institutional framework for their everyday lives, a different social ethos that emerges in the actual living of those lives, and a different social imaginary and non-ideological social (counter-) ideology expressed in their ideas, ideals, aspirations, beliefs, desires, passions and fantasies" (Clark, 37).
14. Michael Tomasello, *The Cultural Origins of Human Cognition* (Cambridge, MA: Harvard University Press, 1999), 79. Early in his own book, Clark refers to psychologist Tomasello's work on human cooperative behavior and mutualism in order to undermine the ideological myth of Hobbesian egoism.
15. Clark, *Impossible Community*, 89.

16   Clark, 89.
17   Clark, 88.
18   This reflects Dussel's point about the performative contradiction within the dominant system. See the previous chapter and "The Architectonic of the Ethics of Liberation," *Philosophy & Social Criticism* 23, no. 3 (1997): 1–35.
19   Clark, *Impossible Community*, 66.
20   "Nature is represented as inessential and massively denied as the unconsidered background to technological society" (Val Plumwood, *Environmental Culture: The Ecological Crisis of Reason* [London: Routledge, 2002], 108).
21   For a brief synopsis of the movement and list of international authors, see the recent book by Michael Löwy, *Ecosocialism: A Radical Alternative to Capitalist Catastrophe* (Chicago, IL: Haymarket Books, 2015), xi–xiv.
22   Joel Kovel, *The Enemy of Nature: The End of Capitalism or the End of the World?*, 2nd ed. (London: Zed Books, 2007), 243. Kovel was a trained physician and psychoanalyst turned political activist, one of the founders of the International Ecosocialist Network, co-author of the Belem Ecosocialist Declaration (2009), and editor of the ecosocialist journal *Capitalism Nature Socialism* from 2003 to 2013.
23   Marxist ecological writer John Bellamy-Foster might dispute this claim, since his groundbreaking book *Marx's Ecology* (New York: Monthly Review Press, 2000) defends the view that Marx himself was quite ecologically sensitive and attuned to the ecological destructiveness as well as the human destructiveness of the capitalist system. Kovel and other ecosocialists think that we can acknowledge the relevance of Marx's critique of capitalism to the ecological crisis without having to argue that Marx himself was an ecological thinker.
24   See his discussions of post-Katrina New Orleans, Zapatistas in Chiapas, Gaviotas Colombia, and South African shack dwellers in Kovel, *Enemy of Nature*, 248–68.
25   Kovel, 214.
26   Joel Kovel, "Ecosocialism, Global Justice, and Climate Change," *Capitalism Nature Socialism*, 19, no. 2 (2008): 9. He writes that "what is called an 'ecocentric ethic' is essentially ethics in defense of intrinsic value. Simply put, it is the refusal to reduce the world to cash, and to knuckle under to the lords of economic calculation" (Kovel, 9). See also the more recent essay "Ecosocialism as a Human Phenomenon," *Capitalism Nature Socialism*, 25, no. 1 (2014): 10–23, where values are again a central theme. For an earlier and more extensive interpretation of Kovel's work in light of my value theory, see my "From Ecological Politics to Intrinsic Value: An Examination of Kovel's Value Theory," *Capitalism Nature Socialism*, 21, no. 3 (2010): 81–101.
27   Kovel, *Enemy of Nature*, 212.
28   As most readers know, this distinction emerges right at the start of Marx's *Capital* in his analysis of the commodity. My discussion relies on those pages. I

should add here that although there is an anthropocentric tendency in Marx's theory (where value arises thanks to human labor of different kinds), Marx leaves room for nonanthropocentric value. Unlabored-upon air, soil, and water are still valuable before humans come along.

29 Cf. Arne Naess's discussion of such properties in *Ecology, Community, Lifestyle* (Cambridge: Cambridge University Press, 1989), 48–51, where he explicitly contrasts them with the subjective and relative, as well as the supposedly objective primary qualities of things.

30 Naess, 213. The social ontology developed by Italian philosopher Maurizio Ferraris in his book *Documentality: Why It Is Necessary to Leave Traces* (New York: Fordham University Press, 2013) gives a good account of the nature of money that is both realist and materialist, but not strictly Marxist. See 154–64.

31 A similar view of the difference of kind between economic and personal values is expressed by O'Neill et al. "An environment matters because it expresses a particular set of relations to one's community that would be betrayed if a price were accepted for it. The treatment of the natural world is expressive of one's attitude to those who passed the land on to you and to those who will follow you. Money is not a neutral measuring rod for comparing the losses and gains in different values. Values cannot all be caught within a monetary currency" (John O'Neill, Alan Holland, and Andrew Light, *Environmental Values* [New York: Routledge, 2010], 79).

32 Kovel, *Enemy of Nature*, 134, 140. See also John Bellamy Foster on the laws of ecology in *The Vulnerable Planet: A Short Economic History of the Environment*, rev. ed. (New York: Monthly Review Press, 1999), 118–24.

33 Kovel, *Enemy of Nature*, 140.

34 Kovel, 214. This reflects the phenomenon of "commodity fetishism" that Clark referred to in the previous section.

35 Kovel, 215.

36 For a history of the concept, see Erik Gómez-Baggethun et al., "The History of Ecosystem Services in Economic Theory and Practice: From Early Notions to Markets and Payment Schemes," *Ecological Economics* 69 (2010): 1209–18.

37 Whether this is a good assumption is still under debate. There is a whole literature on the "biodiversity-ecosystem services" relation alone (related to the longer-standing "diversity-stability" debate). See Kevin deLaplante and Valentin Picasso, "The Biodiversity-Ecosystem Function Debate in Ecology," in *Philosophy of Ecology*, ed. Kevin deLaplante, Bryson Brown, and Kent A. Peacock (Oxford: North-Holland, 2011), 169–200, for a good philosophical introduction.

38 Gómez-Baggethun et al., "History of Ecosystem Services." Timeline appears on 1213.

39 See Joseph Alcamo et al., *Ecosystems and Human Well-Being: Current State and Trends: Findings of the Condition and Trends Working Group of the Millennium*

Ecosystem Assessment, vol. 1 (Washington, DC: Island Press, 2005), and subsequent volumes.

40 Robert Costanza et al., "The Value of the World's Ecosystem Services and Natural Capital," *Nature* 387 (1997): 253–60.

41 For a recent treatment, see Kurt Jax et al., "Ecosystem Services and Ethics," *Ecological Economics* 93 (2013): 260–68.

42 In "Ecosystem Services: From Eye-Opening Metaphor to Complexity Blinder," *Ecological Economics* 69 (2010): 1219–27, Richard Norgaard claims that ES betrays the richness of the ecological sciences. It is not that the ecology is weak, it is that the ecology we do have is very rich, and does not support the ES perspective (1220). Anthropologist Sian Sullivan has also asked fundamental questions about the discourse in "Green Capitalism, and the Cultural Poverty of Constructing Nature as Service Provider," *Radical Anthropology* 3 (2009): 18–27.

43 See especially Noel Castree, "Commodifying What Nature?," *Progress in Human Geography* 27, no. 3 (2003): 273–97; "Neoliberalism and the Biophysical Environment 1: What 'Neoliberalism' Is, and What Difference Nature Makes to It," *Geography Compass* 4, no. 12 (2010): 1725–33, plus another essay in the same series, "Theorising the Neoliberalisation of Nature" (1734–46).

44 Gómez-Baggethun et al., "History of Ecosystem Services."

45 "By the second half of the 20th century land[,] or more generally environmental resources, completely disappeared from the production function and the shift from land and other natural inputs to capital and labor alone, and from physical to monetary and more aggregated measures of capital, was completed." Hubacek and van der Bergh, cited in Gómez-Baggethun et al., "History of Ecosystem Services," 1212. Compare Moore's similar but more developed account of value in *Capitalism in the Web of Life: Ecology and the Accumulation of Capital* (New York: Verso, 2015). "Primary production," or "bio-energy fixation" would be the ultimate extrahuman "unpaid work" upon which capital accumulation depends, in Moore's terms.

46 Gómez-Baggethun et al., "History of Ecosystem Services," 1215.

47 Erik Gómez-Baggethun and Manuel Ruiz-Perez, "Economic Valuation and the Commodification of Ecosystem Services," *Progress in Physical Geography* 35, no. 5 (2011): 621–22.

48 Gómez-Baggethun and Ruiz-Perez, 622.

49 Gómez-Baggethun and Ruiz-Perez, 623. Ecosystem services may be valued, but they need not become part of market exchange on that account.

50 Gómez-Baggethun and Ruiz-Perez, 623.

51 My emphasis. They continue: "Appraisal of valuation cannot be detached from the analysis of the sociopolitical processes through which the market expands its limits and through which economic value colonizes new domains. Monetary valuation of ecosystem services does not equate to commodification of ecosystem

services, but it paves the way (discursively and sometimes technically) for commodification to happen" (Gómez-Baggethun and Ruiz-Perez, 624). They explain their own perspective this way: "We believe that the idea of ecosystem services is a powerful concept that can advance the ontological position that ecosystems are not only a matter of ethics and aesthetics, but also a basic condition for human life and subsistence. Furthermore, we feel that economic valuation can be a potent information tool when not used as a single decision making criteria [sic] (e.g. Cost Benefit Analysis), and if used alongside other valuation methods that capture the non-economic value dimensions of nature.... Our criticism is aimed at the idea that economic valuation can capture a comprehensive picture of nature's societal value and at the belief that economic valuation can solve the problems and shortcomings of traditional conservation" (624).

52 Gómez-Baggethun and Ruiz-Perez, 624.

53 "When exporting market mechanisms for the protection of nature to developing countries and non-market societies, international organizations promoting market mechanisms for conservation can consciously or unconsciously contribute to manufacture the *homo economicus* in places where such logic was inexistent, or culturally discouraged by the existing institutional structures" (Gómez-Baggethun et al., "History of Ecosystem Services," 1216).

54 M. J. Peterson et al., "Obscuring Ecosystem Function with Application of the Ecosystem Services Concept," *Conservation Biology* 24, no. 1 (2009): 116. Internal references removed.

55 Peterson et al., 117.

56 Castree has also concluded that commodification of ecosystem services, just as other forms of neoliberalism, has had negative effects. See his "Neoliberalism and the Biophysical Environment 3: Putting Theory into Practice," *Geography Compass* 5, no. 1 (2011): 42–43.

57 See Val Plumwood, "Being Prey," in *The Ultimate Journey: Inspiring Stories of Living and Dying*, ed. James O'Reilly, Sean O'Reilly, and Richard Sterling (San Francisco, CA: Travelers' Tales, 2000) 128–46. Also available online at http://www.utne.com/arts/being-prey.

58 I am not alone in recommending a shift in collective prioritization like this. Outside of environmental philosophy, Virginia Held also discusses this kind of shifting of collective priorities in relation to militarism, education, and social welfare. "Instead of seeing the corporate sector, and military strength, and government and law as the most important segments of society deserving the highest levels of wealth and power, a caring society might see the tasks of bringing up children, educating its members, meeting the needs of all, achieving peace and treasuring the environment, and doing these in the best ways possible to be that to which the greatest social efforts of all should be devoted" (*The Ethics of Care: Personal, Political, and Global* [Oxford: Oxford University Press, 2006], 19).

59 Max Scheler's critique of Modern morality in the final chapter of his book

Ressentiment (trans. Lewis B. Coser and William W. Holdheim [Milwaukee, WI: Marquette University Press, 2003) provides some valuable insights here.

60   Jonathan Haidt, "The Emotional Dog and Its Rational Tail: A Social Intuitionist Approach to Moral Judgment," *Psychological Review* 108, no. 4 (2001): 822n3.

61   Derek Wall has a valuable recent discussion of the literature in *The Commons in History: Culture, Conflict, and Ecology* (Cambridge, MA: The MIT Press, 2014).

62   Kovel cites this phrase from Marx, but Rousseau had the same idea: "The first man who, having fenced in a piece of land, said 'This is mine,' and found people naive enough to believe him, that man was the true founder of civil society. From how many crimes, wars, and murders, from how many horrors and misfortunes might not any one have saved mankind, by pulling up the stakes, or filling up the ditch, and crying to his fellows: Beware of listening to this impostor; you are undone if you once forget that the fruits of the earth belong to us all, and the earth itself to nobody" (Jean Jacques Rousseau, *On the Origin of the Inequality of Mankind* [1754], chap. 2, https://www.marxists.org/reference/subject/economics/rousseau/inequality/ch02.htm).

63   Wall, *Commons in History*, 131. Kovel claims that "the notion of standing over and against the earth in order to own it is at the core of the domination of nature. A usufructuary is all we can claim with regard to the earth. But this demands that our species proves its worth by using, enjoying, and improving the globe that is our home" (*Enemy of Nature*, 271). While usufruct is technically a "property right," it can easily be defined in processual and relational rather than propertarian terms.

64   Kovel, *Enemy of Nature*, 235–36.

65   Clark highlights the additional important consideration that "one of the basic problems of an oppositional movement is educating the community and propagating transformative values in a context in which the dominant system has a relative monopoly on education and socialization. It is highly unlikely that purely informal and spontaneous efforts at education will allow a movement to proliferate to the point at which it could challenge the dominant system.... The more general lesson that can be learned from this experience is that the quest for the free community will require many dedicated activists who will make it their vocation to develop skills that will help them facilitate the organization of the affinity groups, base communities, and transformed town and neighborhood communities that will form the fabric of the new society" (*Impossible Community*, 229).

66   Inspired by Paolo Freire and informed by varied strands of ethical theory, Enriqué Dussel has perhaps gone further than anyone in describing the process of a community's "coming to consciousness" of their victimization (conscientization), and engaging in critical reasoning that reveals the contradictions in the current system and presents hopeful alternatives. According to Dussel, this new consciousness amounts to a new paradigm that shapes

the value judgments of those involved, aiming "to 'develop' the life of every human subject" in light of the "creative construction of new critical values" (*Ethics of Liberation in the Age of Globalization and Exclusion*, trans. Alejandro A. Vallega et al. [Durham, NC: Duke University Press, 2013], 345). The term *conscientization* comes from the tradition of Latin American liberation philosophy. Enriqué Dussel employs the term in his *Ethics of Liberation* and explains it this way: "The Brazilian neologism *concientizaçao* refers to the 'process of becoming' of a theoretical consciousness ..., and to an ethical consciousness of responsibility.... Semantically, this neologism gives exact expression to the 'conscientization'... that is implicit in the liberation process that is furthered by the affected, the oppressed, and the excluded" (603n23). In other words, the transformation of ethos discussed here involves a consciousness raising that is both intellectual and moral at once, since it keeps the injustice and exploitation of the current arrangements squarely in view, and calls for a response to injustice. See *Ethics of Liberation*, chapters 5 and 6.

CHAPTER SIX

1. *Science as Social Knowledge: Values and Objectivity in Scientific Inquiry* (Princeton, NJ: Princeton University Press, 1990), ix.
2. Richard Levins and Yrjo Haila, *Humanity and Nature: Ecology, Science, and Society* (London: Pluto Press, 1992), 9.
3. Peter Taylor, *Unruly Complexity: Ecology, Interpretation, Engagement* (Chicago, IL: University of Chicago Press, 2005), xiv.
4. Bruno Latour has also addressed this dualism in accounts of knowledge-making in much of his work. See, for example, "One More Turn after the Social Turn," in *The Social Dimensions of Science*, ed. Ernan McMullin (Notre Dame, IN: University of Notre Dame Press, 1992), 272–92.
5. Latour's term for this kind of reasoning is *short-circuiting*. He explains how references to capital-*N* Nature or capital-*S* Science are often used to silence debate about political issues in the first chapter of his *Politics of Nature*, brilliantly using Plato's allegory of the cave as the original model. See his *Politics of Nature: How to Bring the Sciences into Democracy* (Cambridge, MA: Harvard University Press, 2004), 9–52.
6. Broadly speaking, I follow Richard Levins in his recommendation to recognize the dual nature of the scientific enterprise: "Science has a dual nature. On the one hand, it really does enlighten us about our interactions with the rest of the world, producing understanding and guiding our actions.... On the other hand, as a product of human activity, science reflects the conditions of its production and the viewpoints of its producers or owners" (Levins, "Ten Propositions on Science and Antiscience," *Social Text* 46/47 [1996]: 103).
7. Helen Longino, *The Fate of Knowledge* (Princeton, NJ: Princeton University Press, 2002), 8.

8   Longino discusses epistemological and ontological senses of "construction" in *Fate*, 12.
9   Avoiding it in this way results in her own position: "an interdependence interpretation of nonindividualism, a pluralist interpretation of nonmonism, and a contextualist interpretation of nonrelativism." Longino, *Fate*, 89–93. Passages from 92 and 93.
10 Taylor recounts the origin of this difference (*Unruly Complexity*, 56).
11 See Steward Pickett, Jurek Kolasa, and Clive Jones, *Ecological Understanding: The Nature of Theory and the Theory of Nature*, 2nd ed. (New York: Elsevier, 2007), 11.
12 Robert O'Neill et al., *A Hierarchical Concept of Ecosystems* (Princeton, NJ: Princeton University Press, 1986). Ecofeminists Warren and Cheney also discuss hierarchy theory in "Ecological Feminism and Ecosystem Ecology," *Hypatia* 6, no. 1 (1991): 186.
13 O'Neill et al., *Hierarchical Concept*, 7.
14 O'Neill et al., 83.
15 Taylor, *Unruly Complexity*, 233.
16 Pickett, Kolasa, and Jones, *Ecological Understanding*, 151–52.
17 Pickett, Kolasa, and Jones, 153–54. Emphasis in original.
18 They refer to Longino here, though they seriously misunderstand Longino by not grasping the difference between individual bias and systematic community-wide assumptions. See Pickett, Kolasa, and Jones, 178.
19 Taylor, *Unruly Complexity*, xiii.
20 Taylor, xiv, 221.
21 Taylor, 31.
22 Taylor, 102.
23 "Processes of different kinds and scales, involving heterogeneous elements, are interlinked in the production of any outcome and in their own ongoing transformation. Each is implicated in the others" (Taylor, 161).
24 He shares this first step (among others) with developmental systems theorists. See Susan Oyama, Paul E. Griffiths, and Russell Gray, eds., *Cycles of Contingency: Developmental Systems and Evolution* (Cambridge, MA: The MIT Press, 2001), and his contribution to that collection, "Distributed Agency within Intersecting Ecological, Social, and Scientific Processes," 313–32. This is also the basic principle of "flat" ontologies.
25 Taylor, *Unruly Complexity*, 164. "Intermediate complexity accounts *favor the idea of multiple, smaller engagements linked together within the intersecting processes*."
26 Taylor, 164.
27 Taylor, 239.
28 Taylor, 250. "Structuredness is not reducible to micro- or macro-determinations" (Taylor, 163).
29 Taylor, 102, 103. He refers to Marx's concept of labor in this context.

30 "A heterogeneous web of materials, tools, and other people help agents *act as if* the world were like their representations of it" (Taylor, 103).
31 Taylor, 239. In this context, he also criticizes Latour for possessing a narrow conception of motivation, 241.
32 Taylor, 236.
33 Taylor, 237. While he refers specifically to scientific agents here, the same would seem to hold for any human social agent.
34 Taylor, 236.
35 "How Do We Know We Have Global Environmental Problems? Undifferentiated Science-Politics and Its Potential Reconstruction," in *Changing Life: Genomes, Bodies, Ecologies, Commodities*, ed. Peter J. Taylor, Saul Halfon, and Paul Edwards (Minneapolis: University of Minnesota Press, 1997), 149. Philosopher of science Sandra Mitchell also challenges this "predict-and-act" model in *Unsimple Truths: Science, Complexity, and Policy* (Chicago, IL: University of Chicago Press, 2009), chapter 5.
36 Taylor, 97.
37 These relations are cumulative, such that the mental and social depend on the material as well. Dependence may be regarded as transitive, even if determination is not. (Recall the discussion of genetic reductionism in chapter 1.)
38 For more on this distinction, see this author's "Flat, Hierarchical, or Stratified? Determination and Dependence in Social-Natural Ontology," in *New Research on the Philosophy of Nicolai Hartmann*, ed. Keith Peterson and Roberto Poli (Berlin: De Gruyter, 2016), 109–31.
39 "Explanation" is primarily conceived here as a social-communicative, discursive-pragmatic activity of knowledge production oriented by specific epistemic values and practical goals, rather than as a purely rational, "internalist" "mirroring of nature" insulated from the conditions of its production.
40 With the term *flat* or relational ontology I refer to a host of more recent proposals in the domains of science studies, new materialism, neovitalism, posthumanism, and so forth, such as Taylor's view, but also Latour, Jane Bennet, Deleuze and Guatarri, process philosophy, all of which are "categorially" reductive in the above sense.
41 Charles Taylor, "What Is Human Agency?," in *Philosophical Papers I: Human Agency and Language* (Cambridge: Cambridge University Press, 1985), 15–44.
42 Nicolai Hartmann, *Der Aufbau der realen Welt: Grundriß der allgemeinen Kategorienlehre* (Berlin: De Gruyter, 2010), viii.
43 Nicolai Hartmann, *New Ways of Ontology* (Piscataway, NJ: Transaction, 2012), 88.
44 Hartmann, 89–90. There are obvious links between anthropocentrism and all forms of idealism: "In the religious and philosophical worldviews, the anthropocentric interpretation of the world always returns, usually bound up with the devaluation of the real world" (Hartmann, *Das Problem des geistigen*

Seins: Untersuchungen zur Grundlegung der Geschichtsphilosophie und der Geisteswissenschaften [Berlin: W. de Gruyter, 1962], 113).

45 For an analysis of views she calls "nature skepticism" and "nature cynicism," see Val Plumwood, "The Concept of a Cultural Landscape: Nature, Culture and Agency in the Land," *Ethics and the Environment* 11, no. 2 (2006): 115–50.

46 Hartmann, *New Ways of Ontology*, 49.

WORKS CITED

Abram, David. *The Spell of the Sensuous: Perception and Language in a More-Than-Human World*. New York: Vintage, 1997.
Alcamo, Joseph, et al. *Ecosystems and Human Well-Being: Current State and Trends: Findings of the Condition and Trends Working Group of the Millennium Ecosystem Assessment*. Vol. 1. Washington, DC: Island Press, 2005.
Bannon, Bryan. *From Mastery to Mystery: A Phenomenological Foundation for an Environmental Ethic*. Athens: Ohio University Press, 2014.
Beauvoir, Simone de. *The Second Sex*. Translated by H. M. Parshley. New York: Alfred A. Knopf, 1971.
Bender, Frederic. *The Culture of Extinction*. New York: Humanity Books, 2003.
Berry, Thomas. *The Great Work: Our Way into the Future*. New York: Bell Tower, 1999.
Birch, Charles, and John B. Cobb. *The Liberation of Life: From the Cell to the Community*. Cambridge: Cambridge University Press, 1981.
Birch, Thomas H. "Moral Considerability and Universal Consideration." *Environmental Ethics* 15, no. 4 (1993): 313–32.
Bonneuil, Christophe, and Jean-Baptiste Fressoz. *The Shock of the Anthropocene: The Earth, History and Us*. Translated by David Fernbach. London: Verso, 2017.
Bookchin, Murray. *The Philosophy of Social Ecology: Essays on Dialectical Naturalism*. Montreal: Black Rose Books, 1990.
———. "What Is Social Ecology?" In *Environmental Philosophy: From Animal Rights to Radical Ecology*, edited by Michael E. Zimmerman et al., 354–73. Upper Saddle River, NJ: Prentice Hall, 2004.
Braidotti, Rosi. *The Posthuman*. Cambridge: Polity Press, 2013.
Brassier, Ray. "Objects and Concepts." In *The Speculative Turn: Speculative Realisms and Materialisms*, edited by Graham Harman et al., 47–65. Sydney: Re-Press, 2010.

Cadwallader, Eva. "The Main Features of Value Experience." *Journal of Value Inquiry* 14, nos. 3-4 (1980): 229-44.

Callicott, J. Baird. "Intrinsic Value, Quantum Theory, and Environmental Ethics." *Environmental Ethics* 7, no. 3 (1985): 257-75.

———. "The Metaphysical Implications of Ecology." *Environmental Ethics* 8, no. 4 (1986): 301-16.

———. *In Defense of the Land Ethic*. Albany: State University of New York Press, 1989.

———. "Holistic Environmental Ethics and the Problem of Ecofascism." In *Environmental Philosophy: From Animal Rights to Radical Ecology*. 4th ed., edited by Michael Zimmerman et al., 116-29. Upper Saddle River, NJ: Prentice Hall, 2004.

———. "Ecology: An Ethical Perspective." *Nature Education Knowledge* 3, no. 10 (2012): 16-17.

Castoriadis, Cornelius. *The Imaginary Institution of Society*. Translated by K. Blamey. Cambridge: The MIT Press, 1998.

Castree, Noel. "Commodifying What Nature?" *Progress in Human Geography* 27, no. 3 (2003): 273-97.

———. "Neoliberalism and the Biophysical Environment 1: What 'Neoliberalism' Is, and What Difference Nature Makes to It." *Geography Compass* 4, no. 12 (2010): 1725-33.

———. "Theorising the Neoliberalisation of Nature." *Geography Compass* 4, no. 12 (2010): 1734-46.

———. "Neoliberalism and the Biophysical Environment 3: Putting Theory into Practice." *Geography Compass* 5, no. 1 (2011): 35-49.

Chakrabarty, Dipesh. "The Climate of History: Four Theses." *Critical Inquiry* 35, no. 2 (2009): 197-222.

Cheney, Jim. "Eco-Feminism and Deep Ecology." *Environmental Ethics* 9, no. 2 (1987): 115-145.

Chisholm, Roderick M. *Realism and the Background of Phenomenology*. Glencoe, IL: Free Press, 1960.

Clark, John. *The Impossible Community: Realizing Communitarian Anarchism*. New York: Bloomsbury, 2013.

Code, Lorraine. *Ecological Thinking: The Politics of Epistemic Location*. Oxford: Oxford University Press, 2006.

Cosmides, Leda, and John Tooby. "Evolutionary Psychology: A Primer." Center for Evolutionary Psychology. University of California, Santa Barbara. Last modified January 13, 1997. https://www.cep.ucsb.edu/primer.html.

Costanza, Robert, et al. "The Value of the World's Ecosystem Services and Natural Capital." *Nature* 387 (1997): 253-60.

Crist, Eileen. "On the Poverty of Our Nomenclature." *Environmental Humanities* 3 (2013): 129-47.

Cronon, William. "The Trouble with Wilderness; or, Getting Back to the Wrong Nature." In *Uncommon Ground: Rethinking the Human Place in Nature*, edited by William Cronon, 69-90. New York: W. W. Norton, 1995.

deLaplante, Kevin, and Valentin Picasso. "The Biodiversity-Ecosystem Function Debate in Ecology." In *Philosophy of Ecology*, edited by Kevin deLaplante, Bryson Brown, and Kent A. Peacock, 169–200. Oxford: North-Holland, 2011.
Deleuze, Gilles, and Felix Guatarri. *Thousand Plateaus*. Translated by Brian Massumi. Minneapolis: University of Minnesota Press, 1987.
Derrida, Jacques. *The Animal That Therefore I Am*. Translated by David Wills. New York: Fordham University Press, 2009.
Descartes, René. "From the Letters of 1646 and 1649." In *The Philosophical Writings of Descartes*, edited by John Cottingham, Anthony Kenny, Dugald Murdoch, and Robert Stoothoff. Cambridge: Cambridge University Press, 1991.
Dupré, John. *The Disorder of Things: Metaphysical Foundations of the Disunity of Science*. Cambridge: Harvard University Press, 1993.
———. *Human Nature and the Limits of Science*. New York: Oxford University Press, 2002.
———. *Humans and Other Animals*. Oxford: Clarendon, 2002.
Dussel, Enrique. *Ethics of Liberation in the Age of Globalization and Exclusion*. Translated by Alejandro A. Vallega et al. Durham, NC: Duke University Press, 2013.
———. "The Architectonic of the Ethics of Liberation: On Material Ethics and Formal Moralities." *Philosophy Social Criticism* 23, no. 1 (1997): 1–35.
———. "Principles, Mediations, and the 'Good' as Synthesis." *Philosophy Today* 41 (1997): S55–66.
Eldredge, Niles, and Marjorie Grene. *Interactions: The Biological Context of Social Systems*. New York: Columbia University Press, 1992.
Evernden, Neil. *The Natural Alien: Humankind and the Environment*. Toronto: University of Toronto Press, 1993.
Ferraris, Maurizio. *Documentality: Why It Is Necessary to Leave Traces*. Translated by Richard Davies. New York: Fordham University Press, 2013.
———. *Introduction to New Realism*. Translated by Sarah de Sanctis. London: Bloomsbury, 2015.
Findlay, J. N. *Axiological Ethics*. London: Macmillan, 1970.
Foltz, Bruce. "Environmental Continental Philosophy." In *A Dictionary of Continental Philosophy*. New Haven, CT: Yale, 2006.
Foster, John Bellamy. *The Vulnerable Planet: A Short Economic History of the Environment*. Rev. ed. New York: Monthly Review Press, 1999.
———. *Marx's Ecology*. New York: Monthly Review Press, 2000.
Fotopoulos, Takis. "Myths on the Ecological Crisis." *International Journal of INCLUSIVE DEMOCRACY* 3, no. 2 (April 2007). http://www.inclusivedemocracy.org/journal/vol3/vol3_no2_Myths_Ecological_Crisis.htm.
Fox, Warwick. *Toward a Transpersonal Ecology: Developing New Foundations for Environmentalism*. Boston, MA: Shambhala, 1990.
Freud, Sigmund. *The Standard Edition of the Complete Psychological Works of*

Sigmund Freud. Vol. 16. Translated and edited by James Strachey. London: Hogarth Press, 1963.

Gare, Arran. *Postmodernism and the Environmental Crisis*. London: Routledge, 1995.

———. *Nihilism Inc.: Environmental Destruction and the Metaphysics of Sustainability*. Como, NSW, Australia: Eco-Logical Press, 1996.

———. "Philosophical Anthropology, Ethics and Political Philosophy in an Age of Impending Catastrophe." *Cosmos and History: The Journal of Natural and Social Philosophy* 5, no. 2 (2009): 264–86.

Gehlen, Arnold. *Urmensch und Spätkultur: Philosophische Ergebnisse und Aussagen*. Bonn: Athenäum Verlag, 1956.

———. *Man: His Nature and Place in the World*. Translated by Clare McMillan and Karl Pillemer. New York: Columbia University Press, 1988.

Gómez-Baggethun, Erik, Rudolf De Groot, Pedro Lomas, and Carlos Montes. "The History of Ecosystem Services in Economic Theory and Practice: From Early Notions to Markets and Payment Schemes." *Ecological Economics* 69 (2010): 1209–18.

Gómez-Baggethun, Erik, and Manuel Ruiz-Perez. "Economic Valuation and the Commodification of Ecosystem Services." *Progress in Physical Geography* 35, no. 5 (2011): 613–28.

Goodpaster, Kenneth. "On Being Morally Considerable." *Journal of Philosophy* 75, no. 6 (1978): 308–25.

Gould, Stephen Jay. *Ontogeny and Phylogeny*. Cambridge, MA: Harvard University Press, 1977.

———. *Ever Since Darwin*. New York: W. W. Norton, 1992.

Graeber, David. *Toward an Anthropological Theory of Value: The False Coin of Our Own Dreams*. New York: Palgrave, 2001.

Grene, Marjorie. *Approaches to a Philosophical Biology*. New York: Basic Books, 1968.

———. *A Philosophical Testament*. Chicago, IL: Open Court, 1995.

Grey, William. "Anthropocentrism and Deep Ecology." *Australasian Journal of Philosophy* 71, no. 4 (1993): 463–75.

Gruen, Lori. "Refocusing Environmental Ethics: From Intrinsic Value to Endorsable Valuations." *Philosophy & Geography* 5, no. 2 (2002): 153–64.

———. *Entangled Empathy: An Alternative Ethic for our Relationships with Animals*. New York: Lantern Books, 2015.

Guha, Ramachandra. "Radical American Environmentalism and Wilderness Preservation: A Third World Critique." *Environmental Ethics* 11, no. 1 (1989): 71–83.

Habermas, Jürgen. "Discourse Ethics: Notes on a Program of Philosophical Justification." In *Moral Consciousness and Communicative Action*, translated by Christian Lenhardt and Shierry Weber Nicholsen, 43–115. Cambridge, MA: The MIT Press, 1990.

———. "Remarks on Discourse Ethics." In *Justification and Application*, translated by Ciaran Cronin, 19–112. Cambridge, MA: The MIT Press, 1993.
Hacking, Ian. *Representing and Intervening*. Cambridge: Cambridge University Press, 1983.
Haidt, Jonathan. "The Emotional Dog and Its Rational Tail: A Social Intuitionist Approach to Moral Judgment." *Psychological Review* 108, no. 4 (2001): 814–34.
Haraway, Donna. *When Species Meet*. Minneapolis: University of Minnesota Press, 2008.
———. "Anthropocene, Capitalocene, Plantationocene, Chthulucene: Making Kin." *Environmental Humanities* 6 (2015): 159–65.
Harding, Sandra, ed. *The Feminist Standpoint Theory Reader: Intellectual and Political Controversies*. London: Routledge, 2004.
Hargrove, Eugene. "Weak Anthropocentric Intrinsic Value." *Monist* 75, no. 2 (1992): 183–207.
Hartmann, Nicolai. *Der Aufbau der realen Welt: Grundriß der allgemeinen Kategorienlehre*. Berlin: Walter de Gruyter, 1940.
———. *New Ways of Ontology*. Translated by R. Kuhn. Chicago, IL: Henry Regnery, 1953.
———. *Das Problem des geistigen Seins: Untersuchungen zur Grundlegung der Geschichtsphilosophie und der Geisteswissenschaften*. Berlin: Walter de Gruyter, 1962.
———. *Moral Phenomena*. Vol. 1 of *Ethics*. Translated by Stanton Coit. New Brunswick, NJ: Transaction Press, 2002.
———. *Moral Values*. Vol. 2 of *Ethics*. Translated by Stanton Coit. New Brunswick, NJ: Transaction Press, 2003.
———. *Moral Freedom*. Vol. 3 of *Ethics*. Translated by Stanton Coit. New Brunswick, NJ: Transaction Press, 2004.
———. *Aesthetics*. Translated by Eugene Kelly. Berlin: Walter de Gruyter, 2014.
Hedlund-De Witt, A. "Exploring Worldviews and Their Relationships to Sustainable Lifestyles: Towards a New Conceptual and Methodological Approach." *Ecological Economics* 84 (2012): 74–83.
Held, Virginia. *The Ethics of Care: Personal, Political, and Global*. Oxford: Oxford University Press, 2006.
Honneth, Axel, and Hans Joas. *Social Action and Human Nature*. Translated by Raymond Meyer. Cambridge: Cambridge University Press, 1988.
Ingold, Tim. "From Complementarity to Obviation: On Dissolving the Boundaries between Social and Biological Anthropology, Archaeology and Psychology." *Zeitschrift für Ethnologie* 123, no. 1 (1998): 21–52.
Jax, Kurt, et al., "Ecosystem Services and Ethics." *Ecological Economics* 93 (2013): 260–68.
Jensen, Derrick. *Endgame* Vol. 1, *The Problem of Civilization*. New York: Seven Stories Press, 2006.

Joas, Hans. *The Genesis of Values*. Translated by Gregory Moore. Chicago: University of Chicago Press, 2000.

Keller, David R., and Frank B. Golley. *The Philosophy of Ecology: From Science to Synthesis*. Athens: University of Georgia Press, 2000.

Klaver, Irene. "Environmental Continental Philosophy." In *Environmental Philosophy: From Animal Rights to Radical Ecology*, edited by Michael E. Zimmerman et al. Englewood Cliffs, NJ: Prentice Hall, 2004.

Kovel, Joel. *The Enemy of Nature: The End of Capitalism or the End of the World?* 2nd ed. London: Zed Books, 2007.

———. "Ecosocialism, Global Justice, and Climate Change." *Capitalism Nature Socialism* 19, no. 2 (2008): 4–14.

———. "Ecosocialism as a Human Phenomenon." *Capitalism Nature Socialism* 25, no. 1 (2014): 10–23.

Kristeva, Julia. "Women's Time." Translated by Alice Jardine and Harry Blake. *Signs* 7, no. 1 (1981): 13–35.

Laitinen, Arto. *Strong Evaluation without Moral Sources: On Charles Taylor's Philosophical Anthropology and Ethics*. Berlin: Walter de Gruyter, 2008.

Latour, Bruno. "One More Turn after the Social Turn." In *The Social Dimensions of Science*, edited by Ernan McMullin, 272–92. Notre Dame, IN: University of Notre Dame Press, 1992.

———. *Politics of Nature: How to Bring the Sciences into Democracy*. Cambridge, MA: Harvard University Press, 2004.

Lee, Keekok. "The Source and Locus of Intrinsic Value: A Reexamination." *Environmental Ethics* 18, no. 3 (1996): 297–309.

Leopold, Aldo. *A Sand County Almanac: With Essays on Conservation*. New York: Oxford University Press, 1949.

Levins, Richard. "Ten Propositions on Science and Antiscience." *Social Text* 46/47 (1996): 101–11.

Levins, Richard, and Yrjo Haila. *Humanity and Nature: Ecology, Science, and Society*. London: Pluto Press, 1992.

Llewelyn, John. *The Rigor of a Certain Inhumanity: Toward a Wider Suffrage*. Bloomington: Indiana University Press, 2012.

Lloyd, Genevieve. *The Man of Reason*. London: Routledge, 1993.

Longino, Helen. *Science as Social Knowledge: Values and Objectivity in Scientific Inquiry*. Princeton, NJ: Princeton University Press, 1990.

———. *The Fate of Knowledge*. Princeton, NJ: Princeton University Press, 2002.

Löwy, Michael. *Ecosocialism: A Radical Alternative to Capitalist Catastrophe*. Chicago, IL: Haymarket Books, 2015.

Mace, William M. "James J. Gibson's Ecological Approach: Perceiving What Exists." *Ethics & the Environment* 10, no. 2 (2005): 195–216.

Malabou, Catherine. *What Should We Do with Our Brain?* Translated by S. Rand. New York: Fordham University Press, 2008.

Marcuse, Herbert. *One Dimensional Man: Studies in the Ideology of Advanced Industrial Society*. London: Routledge, 2008.

Marx, Karl. *Karl Marx: Selected Writings*. Edited by Lawrence H. Simon. Indianapolis, IN: Hackett, 1994.

Meillassoux, Quentin. "Speculative Realism." With Ray Brassier, Iain Hamilton Grant, and
Graham Harman. *Collapse* 3 (November 2007): 306–449.

Merchant, Carolyn. *The Death of Nature: Women, Ecology, and the Scientific Revolution*. San Francisco, CA: Harper & Row, 1980.

Midgley, Mary. "Duties Concerning Islands." In *Environmental Ethics*, edited by Robert Elliot, 89–103. Oxford: Oxford University Press, 1995.

———. *The Ethical Primate: Humans, Freedom, and Morality*. New York: Routledge, 1996.

Mitchell, Sandra. *Unsimple Truths: Science, Complexity, and Policy*. Chicago, IL: University of Chicago Press, 2009.

Moore, Jason. *Capitalism in the Web of Life: Ecology and the Accumulation of Capital*. New York: Verso, 2015.

Naess, Arne. "The Shallow and the Deep, Long-Range Ecology Movement: A Summary." *Inquiry: An Interdisciplinary Journal of Philosophy and the Social Sciences* 16, nos. 1–4 (1973): 95–100.

———. "The World of Concrete Contents." *Inquiry: An Interdisciplinary Journal of Philosophy and the Social Sciences* 28 (1985): 417–28.

———. "The Deep Ecology Movement: Some Philosophical Aspects." In *Deep Ecology for the 21st Century*, edited by George Sessions, 64–84. Boston, MA: Shambhala, 1995.

———. *Ecology, Community, and Lifestyle: Outline of an Ecosophy*. Translated and edited by David Rothenberg. Cambridge: Cambridge University Press, 1989.

Norgaard, Richard. "Ecosystem Services: From Eye-Opening Metaphor to Complexity Blinder." *Ecological Economics* 69 (2010): 1219–27.

Norton, Bryan G. *Toward Unity among Environmentalists*. New York: Oxford University Press, 1991.

———. "Epistemology and Environmental Values." *Monist* 75, no. 2 (1992): 208–26.

O'Neill, John. "The Varieties of Intrinsic Value." *Monist* 75, no. 2 (1992): 119–37.

———. "Meta-Ethics." In *Companion to Environmental Philosophy*, edited by Dale Jamieson, 163–76. Malden, MA: Blackwell, 2001.

O'Neill, John, Alan Holland, and Andrew Light. *Environmental Values*. New York: Routledge, 2010.

O'Neill, Robert et al. *A Hierarchical Concept of Ecosystems*. Princeton, NJ: Princeton University Press, 1986.

Orr, David. *Ecological Literacy*. Albany: State University of New York Press, 1992.

Ouderkirk, Wayne, and Jim Hill, eds. *Land, Value, and Community: Callicott and Environmental Philosophy*. Albany: State University of New York Press, 2002.

Oyama, Susan, Paul Griffiths, and R. D. Gray. *Cycles of Contingency: Developmental Systems and Evolution*. Cambridge, MA: The MIT Press, 2001.

Passmore, John. *Man's Responsibility for Nature*. 2nd ed. New York: Scribner, 1974.

Peterson, Keith R. "All That We Are: Philosophical Anthropology and Ecophilosophy." *Cosmos and History: The Journal of Natural and Social Philosophy* 6, no. 1 (2010): 60–82.

———. "From Ecological Politics to Intrinsic Value: An Examination of Kovel's Value Theory." *Capitalism Nature Socialism* 21, no. 3 (2010): 81–101.

———. "Bringing Values Down to Earth: Max Scheler and Environmental Philosophy." *Appraisal: The Journal of the Society for Post-Critical and Personality Studies, Re-Appraisal: Max Scheler (Pt 2)* 8, no. 4 (2011): 3–12.

———. "Flat, Hierarchical, or Stratified? Determination and Dependence in Social-Natural Ontology." In *New Research on the Philosophy of Nicolai Hartmann*, eds. Keith Peterson and Roberto Poli, 109–31. Berlin: De Gruyter, 2016.

Peterson, M. J., D. M. Hall, A. M. Feldpausch-Parker, and T. R. Peterson. "Obscuring Ecosystem Function with Application of the Ecosystem Services Concept." *Conservation Biology* 24, no. 1 (2009): 113–19.

Pickett, Steward, Jurek Kolasa, and Clive Jones. *Ecological Understanding: The Nature of Theory and the Theory of Nature*. 2nd ed. New York: Elsevier, 2007.

Plumwood, Val. "Nature, Self, and Gender: Feminism, Environmental Philosophy, and the Critique of Rationalism." *Hypatia* 6, no. 1 (1991): 3–27.

———. *Feminism and the Mastery of Nature*. London: Routledge, 1993.

———. "The Politics of Reason: Towards a Feminist Logic." *Australasian Journal of Philosophy* 71, no. 4 (1993): 436–62.

———. "Androcentrism and Anthrocentrism: Parallels and Politics." *Ethics and the Environment* 1, no. 2 (1996): 119–52.

———. "Being Prey." In *The Ultimate Journey: Inspiring Stories of Living and Dying*, edited by James O'Reilly, Sean O'Reilly, and Richard Sterling, 128–46. San Francisco, CA: Travelers' Tales, 2000.

———. *Environmental Culture: The Ecological Crisis of Reason*. London: Routledge, 2002.

———. "The Concept of a Cultural Landscape: Nature, Culture and Agency in the Land." *Ethics and the Environment* 11, no. 2 (2006): 115–50.

Protevi, John. *Political Affect: Connecting the Social and the Somatic*. Minneapolis: University of Minnesota Press, 2009.

Rees, William. "Human Nature, Eco-Footprints, and Environmental Injustice." *Local Environment* 13, no. 8 (2008): 685–701.

———. "What's Blocking Sustainability? Human Nature, Cognition, and Denial." *Sustainability: Science, Practice, & Policy* 6, no. 2 (2010): 13–25.

Regan, Tom. "Animal Rights, Human Wrongs." *Environmental Ethics* 2, no. 2 (1980): 99–120.

———. "Does Environmental Ethics Rest on a Mistake?" *Monist* 75, no. 2 (1992): 161–82.

Robbins, Paul. *Political Ecology: A Critical Introduction*. 2nd ed. Malden, MA: Wiley-Blackwell, 2012.

Rolston, Holmes, III. *Philosophy Gone Wild: Essays in Environmental Ethics*. Blue Ridge Summit, PA: Prometheus Books, 1986.

———. "Value in Nature and the Nature of Value." In *Philosophy and the Natural Environment*, edited by Robin Attfield and Andrew Belsey, 13–30. Cambridge: Cambridge University Press, 1994.

———. "Environmental Virtue Ethics: Half the Truth but Dangerous as a Whole." In *Environmental Virtue Ethics*, eds. Ronald Sandler and Philip Cafaro, 61–78. Lanham, MD: Rowman & Littlefield, 2005.

———. *A New Environmental Ethics: The Next Millennium of Life on Earth*. London: Routledge, 2012.

Rousseau, Jean Jacques. "On the Origin of the Inequality of Mankind." 1754. Marxists.org. Accessed June 6, 2014. https://www.marxists.org/reference/subject/economics/rousseau/inequality/ch02.htm.

Routley, Richard, and Val Routley. "Against the Inevitability of Human Chauvinism." In *Ethics and Problems of the 21st Century*, edited by K. E. Goodpaster and K. M. Sayre, 36–59. Notre Dame, IN: University of Notre Dame Press, 1979.

———. "Social Theories, Self-Management, and Environmental Problems." In *Environmental Philosophy*, eds. Don Mannison, Michael McRobbie, and Richard Routley, 217–331. Canberra: Australian National University, 1980.

Rowlands, Mark. *The Environmental Crisis: Understanding the Value of Nature*. New York: St. Martin's, 2000.

Sale, Kirkpatrick. *After Eden: The Evolution of Human Domination*. Durham, NC: Duke University Press, 2006.

Salminen, Antti, and Tere Vadén. *Energy and Experience: An Essay in Nafthology*. Chicago, IL: MCM, 2015.

Sartre, Jean-Paul. "Existentialism Is a Humanism." In *Existentialism from Dostoyevsky to Sartre*, edited by Walter Kaufman. Translated by Philip Mairet. New York: Meridian, 1989.

Scheler, Max. *Philosophical Perspectives*. Translated by O. Haac. Boston, MA: Beacon Press, 1958.

———. *Formalism in Ethics and Non-formal Ethics of Values*. Translated by Manfred S. Frings and Roger L. Funk. Evanston, IL: Northwestern University Press, 1973.

———. *Ressentiment*. Translated by Lewis B. Coser. Milwaukee, WI: Marquette University Press, 1994.

———. *The Human Place in the Cosmos*. Translated by Manfred S. Frings. Evanston, IL: Northwestern University Press, 2009.

Shepard, Paul. *Coming Home to the Pleistocene*. Washington, DC: Island Press, 2004.

Shiva, Vandana. "The Impoverishment of the Environment: Women and Children Last." In *Ecofeminism*, eds. Maria Mies and Vandara Shiva, 70–90. Atlantic Highlands, NJ: Zed Books, 1993.

Shrader-Frechette, K. S., and Earl D. McCoy. *Method in Ecology: Strategies for Conservation.* Cambridge: Cambridge University Press, 1993.

Singer, Peter. *Applied Ethics.* Oxford: Oxford University Press, 1986.

Soper, Kate. *What Is Nature?* Cambridge, MA: Blackwell, 1995.

Stepan, Nancy Leys. "Race and Gender: The Role of Analogy in Science." *Isis* 77, no. 2 (1986): 261–77.

Stone, Christopher. "Moral Pluralism and the Course of Environmental Ethics." In *Environmental Ethics: An Anthology,* edited by Holmes Rolston III and Andrew Light, 193–202. Malden, MA: Blackwell, 2003.

Sullivan, Sian. "Green Capitalism, and the Cultural Poverty of Constructing Nature as Service Provider." *Radical Anthropology* 3 (2009): 18–27.

Szeman, Imre, and Dominic Boyer. *Energy Humanities: An Anthology.* Baltimore, MD: Johns Hopkins University Press, 2017.

Taylor, Charles. *Human Agency and Language: Philosophical Papers 1.* Cambridge: Cambridge University Press, 1985.

Taylor, Paul W. *Respect for Nature: A Theory of Environmental Ethics.* Princeton, NJ: Princeton University Press, 1986.

———. "The Ethics of Respect for Nature." In *Environmental Ethics: An Anthology,* edited by Andrew Light and Holmes Rolston III, 74–84. Oxford: Blackwell Publishing, 2003.

Taylor, Peter. *Unruly Complexity: Ecology, Interpretation, Engagement.* Chicago, IL: University of Chicago Press, 2005.

———. "How Do We Know We Have Global Environmental Problems? Undifferentiated Science-Politics and Its Potential Reconstruction." In *Changing Life: Genomes, Bodies, Ecologies, Commodities,* edited by Peter J. Taylor, Saul Halfon, and Paul Edwards, 149–74. Minneapolis: University of Minnesota Press, 1997.

Toadvine, Ted, and Charles S. Brown, eds. *Eco-Phenomenology: Back to the Earth Itself.* Albany: State University of New York Press, 2003.

Tomasello, Michael. *The Cultural Origins of Human Cognition.* Cambridge, MA: Harvard University Press, 1999.

Treanor, Brian. *Emplotting Virtue: A Narrative Approach to Environmental Virtue Ethics.* Albany: State University of New York Press, 2014.

Varner, Gary. "Can Animal Rights Activists be Environmentalists?" In *Environmental Philosophy and Environmental Activism,* edited by Don Marietta and Lester Embree, 169–201. Lanham, MD: Rowman & Littlefield, 1995.

Wall, Derek. *The Commons in History: Culture, Conflict, and Ecology.* Cambridge, MA: The MIT Press, 2014.

Warren, Karen. "The Power and Promise of Ecological Feminism." *Environmental Ethics* 12 (1990): 126–46.

———. "The Power and Promise of Ecofeminism Revisited." In *Environmental*

Philosophy: From Animal Rights to Radical Ecology, edited by Michael Zimmerman et al., 252–79. Upper Saddle River, NJ: Prentice Hall, 2004.

Warren, Karen, and Jim Cheney. "Ecological Feminism and Ecosystem Ecology." *Hypatia* 6, no. 1 (1991): 179–97.

Watts, Michael J. "Political Ecology." In *A Companion to Economic Geography*, edited by E. Sheppard and T. Barnes, 257–75. Oxford: Blackwell, 2000.

Wenz, Peter. *Environmental Justice*. Albany: State University of New York Press, 1988.

Weston, Anthony. "Beyond Intrinsic Value: Pragmatism in Environmental Ethics." *Environmental Ethics* 7, no. 4 (1985): 321–39.

———. "Between Means and Ends." *Monist* 75, no. 2 (1992): 236–49.

———. "Multicentrism: A Manifesto." *Environmental Ethics* 26, no. 1 (2004): 25–40.

———. *The Incompleat Eco-philosopher*. Albany: State University of New York Press, 2009.

Wexler, Bruce. *Brain and Culture: Neurobiology, Ideology, and Social Change*. Cambridge, MA: The MIT Press, 2006.

White, Lynn, Jr. "The Historical Roots of our Ecologic Crisis." *Science* 155, no. 3767 (1967): 1203–7.

Zerzan, John. *Against Civilization: Readings and Reflections*. Port Townsend, WA: Feral House, 2005.

INDEX

action, 14, 24, 30, 36, 51, 53, 56–60, 62–65, 67, 69–70, 79, 83, 85–86, 89, 97, 99–100, 102–3, 105–6, 109–10, 113, 115, 140, 154, 156–57, 159–62, 163, 165, 186n27, 187n46, 187n50
affordance, 58–59, 61–62, 66–67, 69–70, 84, 86, 187n39
anthropocentrism, 1–2, 7–9, 13–14, 19–27, 32–33, 37, 39–40, 48–49, 73–74, 87, 90, 111, 115, 145, 166, 172n6, 178n16, 189n4, 192n41, 205n44
articulation, 63, 67–68, 83, 88–91, 95, 189n67, 192n43
axiology, 2–3, 159

Callicott, J. Baird, 2–3, 75–76, 108–9
capitalism, 5, 83, 119, 125–29, 136, 139–40, 154–55, 159, 167, 174
Clark, John P., xi, xiv, 12, 117, 119–24, 127, 129
Commons, 132, 137–39

dependence, asymmetrical, ix, 7, 12, 14, 24, 32, 39, 141, 146, 149, 163, 168, 183n2
  axiological, 103, 108, 110, 135

denial of, 7, 28, 30, 32, 40, 133–35, 165, 167–68
ontological, 103, 105–6, 108–09, 111
principle of, ix, 7, 13–14, 28, 39, 70, 109, 140, 146, 149, 159, 166, 169, 171–75
dualism, x, 3, 6–8, 10, 12–15, 19–20, 23, 25–27, 29, 31, 33–34, 37, 40–41, 43, 46–49, 53, 56–57, 60, 62–64, 69–70, 75, 77–78, 80–81, 92, 101, 110, 118, 129, 133, 146–56, 159–60, 166, 175, 181n62, 183n1, 203n4
Dupré, John, 44–45

eccentric stratum, 165–66
ecofeminism, 6, 8, 9, 12, 13, 24, 91, 117, 158
ecological materialism, 12, 15, 47–48, 61
ecosocialism, 117, 125–26
ecosystem services, 29, 83, 119, 125, 130–34, 137, 146, 197, 201
embeddedness, 12, 14–15, 31, 44, 47, 49–52, 63, 70, 74, 102–3, 124–25, 139, 141, 157, 160, 173
environmental justice, 9, 100, 114, 117
ethos, 12, 15, 45, 83–84, 105, 107, 110, 112–13, 117–22, 125, 135–37, 140, 159, 173–74, 197n13, 202n66

223

Evernden, Neil, 37–39
evolutionary psychology, 41, 44, 184n8

Gare, Arran, 3, 19, 35niv, 87, 122
Gehlen, Arnold, 53–58, 60–64, 67, 70, 184n13, 188n60
Gould, Stephen J., 53–55, 185n22
Grene, Marjorie, x, 48, 51, 53–54, 56, 58, 62, 185n22

Kovel, Joel, xi, 12, 19, 119, 125–29, 132, 138, 140, 198n23, 202n63

logic of domination, 6–7, 25–26, 120, 129

mechanism, 1, 3, 7, 10–11, 38–39, 44, 91, 155, 162, 167, 174
metaethics, 4, 14, 48, 64, 73, 93
metascientific stance, xi, 11–16, 140–41, 145–49, 151, 154, 157–61, 165, 174–75

naturalism, 12, 29, 31, 41, 43, 45, 48–49, 61
neoteny, 45, 54–55, 182n73, 185n17
Nietzsche, Friedrich, 35nii, 53–54, 58
normativity, 69, 93

O'Neill, John, 74, 76, 88
ontogeny, 12, 45, 47, 49, 51–52, 58, 61–62, 64, 70, 74, 98–99, 102, 134, 172
ontogeny principle, 98–99, 102
ontology, ix, xi, 3–5, 7, 10, 12–13, 15, 25, 30, 39, 74, 78, 81, 101–2, 140, 146, 149, 159–60, 162, 164–65, 166–69, 174–75, 198n30, 205n40

phenomenology, 4, 37–39, 59, 85, 94, 181n61, 191n24
philosophical anthropology, ix, x, 4, 14, 19, 25, 27, 31–33, 40, 45–48, 58, 69, 148, 160, 169, 172, 183n1
phylogeny, 44–45, 51, 185n22
Plumwood, Val, x, 6–7, 12, 14, 19–20, 22, 24–26, 28–31, 36, 38–39, 47–48, 62, 69, 84–85, 87, 90–92, 101, 110–11, 118, 120, 122, 125, 129, 150
pluralism, 92, 97, 123, 154, 158, 160, 167
political ecology, 1, 12, 15, 30, 115, 117–18, 120, 125, 127, 146, 158–60, 167–69, 175
Portmann, Adolf, 55–56
posthuman, 31–32, 179n22, 205n40
primitivism, 34, 40–41, 45, 53, 172
priorities, 4, 15, 45, 66, 79, 81, 83–84, 87–90, 92, 97, 106, 109, 115, 128, 137, 140, 156, 158, 160, 168, 173–74, 188n66
propertarian, 80–85, 91, 93, 101, 108, 114–15, 119, 129–30, 132, 136–37, 173, 191n24, 202n63

realism, 12, 29, 47, 150, 154, 166, 179n20, 181n62
reductionism, 45, 48, 63, 161, 166, 181n60, 183n81, 205n37
reflexivity, 147, 151–53, 159
relativism, 3, 10, 12, 66, 76–78, 97–98, 100, 102, 104, 109, 203n9
relativist principle, 97–98, 100, 102, 104
relief, 53–54, 58, 60, 70, 80, 84, 86, 89, 97, 186n27
Rolston, Holmes, 75–77, 98, 102, 104, 115, 193n50, 193n56
Rolston's principle, 98, 102, 104

Scheler, Max, x, 32, 35nii, 93, 186n27, 189n71, 189n1, 193n56, 194n3, 201n59
Shepard, Paul, 41–44
social ecology, 4, 117, 120
stratification, 149, 160–61, 166–68
superposition, 161, 165, 168
survival principle, 98–99, 102, 105, 194n1
Taylor, Charles, 68, 93
Taylor, Paul W., 22, 33, 36–37, 39, 63, 77, 111–12
Taylor, Peter J., 12, 153–59, 205n40
Tomasello, Michael, 48, 51–52, 61–62, 66–67, 70, 88, 121, 184n8

value, economic, 73, 108, 119, 123, 125, 127–28, 131–32, 134, 200n51
   goods, 68, 92, 99–100, 103–5, 108, 110–12, 114, 123–25, 127–28, 137, 148, 173, 194n6, 195n12, 196n20, 196n22
   instrumental, 74, 76–78, 112, 128, 173, 191n25
   intrinsic, x, 1–3, 10, 13–14, 65, 69, 70, 73–75, 77–81, 85, 89, 97, 100, 104–5, 112, 126–27, 129, 131, 135, 145, 172, 177n8, 186n30, 196n26
   moral, 89, 92, 99–106, 108–14, 118–19, 123–24, 127, 137–38, 140, 148, 158–59, 168, 173–74, 194n3, 194n4, 195n12, 196n22
   vital, 98, 101–4, 107–8, 110–14, 119, 123–25, 130, 134–35, 137–40, 158, 165, 174, 175

value theory, ix, xi, 2–4, 7, 10, 12, 13, 15, 25, 48, 59, 63, 73–74, 77, 83–85, 92–93, 97, 99–100, 109, 119, 121–22, 136, 141, 146, 148, 158, 160, 169, 173, 188n55, 193n53, 195n4, 195n10

worldview, 1, 3–4, 10–15, 33, 37–39, 140–41, 145–48, 152, 155, 157–61, 166, 169, 172, 174–75, 205n44

www.ingramcontent.com/pod-product-compliance
Ingram Content Group UK Ltd.
Pitfield, Milton Keynes, MK11 3LW, UK
UKHW041917140426
5217IPUK00013B/204